AF553827

Biochemistry

Biochemistry

By

S.A. Iqbal
Y. Mido

DISCOVERY PUBLISHING HOUSE
NEW DELHI

First Published – 2005

Reprinted – 2024

ISBN: 978-81-7141-985-2

Biochemistry

Published by:

DISCOVERY PUBLISHING HOUSE

4383/4B, Ansari Road, Darya Ganj

New Delhi-110 002 (India)

Phone: +91-11-23279245; 23253475; 43596065

Mobile: +91 9811179893 / +91 9871656464

E-mail: discoverybooksindia@gmail.com

orderdphbooks@gmail.com

namitwasan9@gmail.com

web: www.discoverypublishinggroup.com

Printed at:

Infinity Imaging Systems

Delhi

Preface

Biochemistry is considered as on essential subject for students of the biomedical sciences. The book is written to be self-contained and is directed towards students of biomedical disciplines, which demand knowledge of molecular interactions and concepts.

It is hoped that the reader will feel capable of development within his/her chosen specialist subject from a position of molecular strength and thereby be comfortable in the reading of contemporary texts.

This book is very much useful for the B.Sc. and M.Sc. students of Biochemistry.

I am extremely grateful to Mr. Tilak Raj the Prof. of Discovery Publishing House for his helpful advice.

—Authors

Contents

Oxidative Metabolism

1.1 Energy Rich Components

During complete stepwise oxidation of glucose, a primary metabolic fuel in cells, a large quantity of energy is available. The free energy released during the oxidation of glucose in a functioning cells is illustrated in the following equation:

$$C_2H_{12}O_6 + 6\ O_2 \rightarrow 6\ CO_2 + 6\ H_2O \qquad \Delta G^o = -686{,}000 \text{ cal mol}^{-1}$$

When this process occurs under aerobic conditions in most cells, it is possible to conserve less than one half of this "available" energy as 38 molecules of ATP. The ΔG^o values for oxidation of other metabolic fuels are listed in Table 1.1 Carbohydrates and proteins (amino acids) have a caloric value of 3-4 kcal g^{-1}, while lipid (i.e., palmitate, a long-chain fatty acid, or a triacylglycerol) exhibits a caloric value nearly three times greater. The reason that more energy can be derived from lipid than from carbohydrate or protein relates to the average oxidation state of the carbon atoms in these substances. Carbon atoms in carbohydrates are considerably more oxidized (or less reduced) than those in lipids. Hence during sequential breakdown of lipid nearly three times as many **reducing equivalents** (a reducing equivalent is defined as a proton plus an electron, i.e., $H^+ + e^-$) can be extracted than from carbohydrate. Reducing equivalents may be utilized for ATP synthesis in the mitochondrial energy transduction sequence.

The two terminal phosphoryl groups of ATP contain energy-rich or high-energy bonds. What this description is intended to convey is that the free energy of hydrolysis of an energy-rich phosphoanhydriate bond is much greater than would be obtained for a simple phosphate ester. High-energy is not synonymous with stability of the bonding arrangement in question, nor does it refer to the energy required to break such bonds. The concept of high-energy compounds implies that the products of the hydrolytic cleavage of the energy-rich bond are in more stable forms than the original compounds. As a rule, simple phosphate esters (low-energy compounds) exhibit negative ΔG° values of hydrolysis in the range 1-3 kcal mol^{-1}, whereas high-energy bonds have negative ΔG° values in the range 5-15 kcal mol^{-1}. Phosphate esters such as glucose 6-phosphate and glycerol 3-phosphate are examples of low-energy compounds. Various types of energy-rich compounds with approximate values for their ΔG° values of hydrolysis.

There are various reasons why certain compounds of bonding arrangements are energy rich.

Free-Energy Changes and Caloric Values Associated with the Total Metabolism of Various Metabolic Fuels

Compound	*Molecular Weight*	*ΔG° (kcal mol^{-1})*	*Caloric value (kcal g^{-1})*
Glucose	180	–686	3.81
Lactate	90	–326	3.62
Palmitate	256	–2380	9.30
Tripalmitin	809	–7510	9.30
Glycine	75	–234	3.12

First, products of the hydrolysis of an energy-rich bond may exist in more **resonance forms** than the precursor molecule. The more possible resonance forms in which a molecule can exist stabilize that molecule.

Fewer resonance forms can be written for ATP or **pyrophosphate** (PP_i) than for phosphate (P_i).

Second, many high-energy bonding arrangements have groups of similar electrostatic charges located in close proximity to each other in such compounds. Because like charges repel one another, hydrolysis of energy-rich bonds alleviates this situation and, again, lends stability to the products of hydrolysis. Third, hydrolysis of certain energy-rich bonds results in the formation of an unstable compounds, which may isomerize spontaneously to form a more stable compound. Hydrolysis of phosphoenolpyruvate is an example of this type of compound. The ΔG° for isomerization is considerable, and the final product, in this case pyruvate, is much more stable. Finally, if a product of the hydrolysis of a high-energy bond is an undissociated acid, dissociation of the proton and its subsequent buffering may contribute to the overall ΔG° of the hydrolytic reaction. In general, any property or process that lends stability to products of hydrolysis tends to confer a high-energy character to that compounds.

The high-energy character of 3′, 5′-**cyclic adenosine monophosphate (cAMP)** has been attributed to the fact that the phosphoanhydride bonding character in this compound is strained as it bridges the 3′ and 5′ positions on the ribose. The energy-rich character of thiol ester compounds such as acetyl CoA or succinyl CoA results from the relatively acidic character of the thiol group. Hence **acetyl CoA** is nearly equivalent to an anhydride rather than a simple thioester.

The ΔG° value of hydrolysis of the terminal phosphate of ATP is difficult to determine by simply utilizing the K_{eq} of the hydrolytic reaction because of the position of the equilibrium.

$$\text{ATP} + \text{HOH} \rightleftharpoons \text{ADP} + \text{P}_i + \text{H}^+$$

However, the ΔG° of hydrolysis of ATP can be determined indirectly because of the additive nature of free-energy changes. Hence free energy of hydrolysis of ATP can be determined by adding ΔG° of an ATP-utilizing reaction such as hexokinase to ΔG° of a reaction that cleaves the phosphate from the product of the hexokinase reaction, glucose 6-phosphate.

Energy-rich compounds can transfer various groups from the parent (donor) compound to an acceptor compound in a

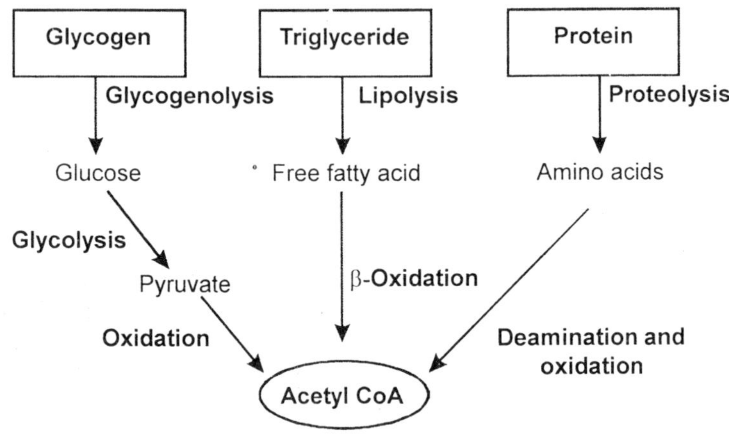

General precursors of acetyl CoA.

thermodynamically feasible fashion as long as an appropriate enzyme is present to facilitate the transfer. The energy-rich intermediates in the glycolytic pathway such as **1,3-bisphosphoglycerate** and phosphoenolpyruvate can transfer their high-energy phosphate

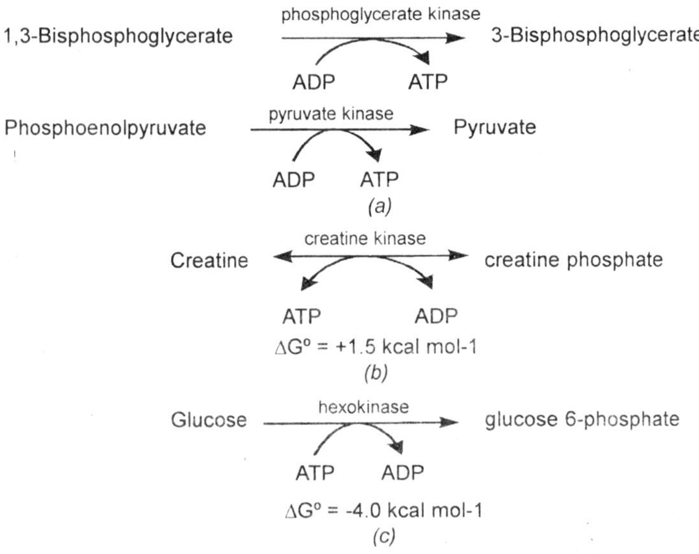

Examples of reaction's involved in transfer of "high-energy" phosphate.

moieties to ATP in the **phosphoglycerate kinase** and **pyruvate kinase** reactions, respectively. The ΔG^{o} values of these two reactions are –4.5 and –7.5 kcal mol^{-1}, respectively, and hence transfer of "high-energy" phosphate is thermodynamically possible, and ATP synthesis is the result. ATP can transfer its terminal high-energy phosphoryl groups to form compounds of relatively similar high-energy character [i.e., **creatine phosphate** in the **creatine kinase** reaction or compounds that are of considerably lower energy, such as glucose 6-phosphate formed in the hexokinase reaction.

Thus phosphate or other transferable groups can be transferred from compounds that contain energy rich bonding arrangements to compounds that have bonding characteristics of a lower energy in thermodynamically permissible enzymatic reactions. This principle is a major premise of the interaction between energy-producing and energy-utilizing metabolic pathways in living cells.

1.2 Pyruvate Dehydrogenase Reaction

During aerobic glycolysis glucose or other monosaccharides are converted to pyruvate, the end product of this cytosolic pathway. Also, degradation of amino acids such as alanine, serine, and cysteine results in the production of **pyruvate**. Pyruvate has a variety of metabolic fates, depending on the tissues and the

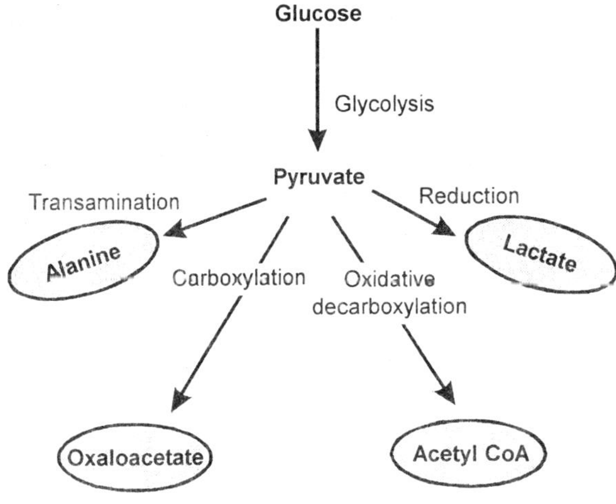

Metabolic fates of pyruvate.

metabolic state of that tissue. The oxidative decarboxylation of pyruvate in the **pyruvate dehydrogenase** reaction is discussed next.

Pyruvate is converted to acetyl CoA by the pyruvate dehydrogenase multienzyme complex.

$$\text{Pyruvate} + NAD^{+} + \text{CoASH} \xrightarrow{\text{pyruvate dehydrogenase}}$$
$$\text{acetyl CoA} + CO_2 + NADH + H^{+} \quad \Delta G^{\circ} = -8 \text{ kcal mol}^{-1}$$

This enzyme is located exclusively in the mitochondrial matrix and is present in high concentrations in tissues such as cardiac muscle and kidney. Because of the large negative ΔG° of the pyruvate dehydrogenase reaction, under physiological conditions the reaction is irreversible. This fact is the primary reason that a net conversion of fatty acid carbon to carbohydrate cannot occur; for example, acetyl CoA from fatty acids cannot be converted to pyruvate. Molecular weights of the multienzyme complex derived from kidney, heart, or liver range from 7 to 8.5 × 10^6. The

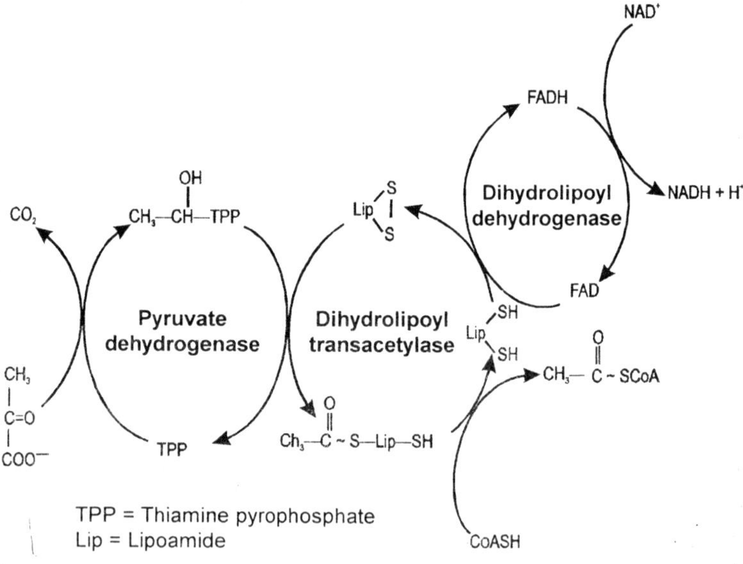

Mechanisms of the pyruvate dehydrogenase reaction; the pyruvate dehydrogenase multienzyme complex.

mammalian pyruvate dehydrogenase enzyme complex consists of three different types of catalytic subunits.

1.3 Energy-utilizing Reactions

Living cells are composed of a complex, intricately regulated system of energy-producing and energy-utilizing chemical reactions. Metabolic reactions involved in energy generation break down ingested or stored fuels such as carbohydrate, lipid, or protein in what are termed **catabolic** pathways. These reactions usually result in the conversion of large complex molecules to smaller molecules (ultimately CO_2 and H_2O), with production of storable or conservable energy, and often require the consumption of oxygen during this process. Such reactions are accelerated during periods of fuel deprivation or stress to an organism.

Energy-utilizing reactions perform various necessary, and in many instances tissue-specific, cellular functions, for example, nerve impulse conduction, muscle contraction, growth, and cell division. Metabolic pathways involved in the biosynthesis of large, complex molecules from smaller precursors are termed **anabolic** pathways and require the expenditure of energy. Such reactions are accelerated when energy is readily available, when precursor molecules are in abundance, or during periods of growth or regeneration of cellular material.

Energy may be derived from oxidation of metabolic fuels presented to the organism usually in the form of carbohydrate, lipid, and protein. The proportion of each fuel utilized as an energy source depends on the tissue and the dietary and hormonal state of the organism. For example, mature erythrocytes and adult brain in the fed state use only carbohydrate as a source of energy, where the liver of a diabetic or fasted individual metabolizes primarily lipid to meet the energy demands. Energy may be consumed during performance of various energy-linked (work) functions. Note that the liver and the pancreas are primarily involved in the biosynthetic and secretory work functions, whereas the primary function of cardiac and skeletal muscle involves converting metabolic energy into mechanical energy during muscle contraction.

The essential link between energy-producing and energy-utilizing pathways is the nucleoside triphosphate, **adenosine 5′-triphosphate (ATP)**. The ATP molecule is a purine (adenine) nucleotide in which the adenine is attached in a glycosidic linkage

Adenosine 5′-triphosphate **Adenosine 5′-triphosphate**

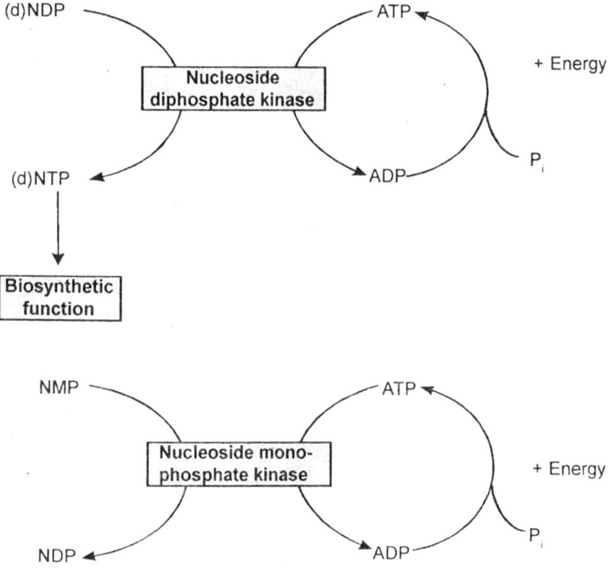

Nucleoside diphosphate kinase and nucleoside monophosphate kinase reactions.

to D-ribose. Three phosphoryl groups are esterified to the 5 position of the ribose moiety in **phosphoanhydride bonds.** The two terminal phosphoryl groups (i.e., β and γ) are involved in the phosphoric acid anhydride bonding and are designated as energy-rich or **high-energy bonds.** Synthesizing ATP as a result of a catabolic process or consuming ATP in an energy-linked cellular function involves formation and either hydrolysis or transfer of the terminal phosphate group of ATP. The physiological form of this nucleotide is chelated with a divalent metal cation such as magnesium. Adenosine diphosphate also chelates magnesium, but the affinity of the metal cation for ADP is considerably less than for ATP. Although adenine nucleotides are mainly involved in energy generation or conservation, various nucleoside triphosphates, including ATP, are involved in transferring energy during bio-synthetic processes. The guanine nucleotide **GTP** serves as the source of energy in gluconeogenesis and protein synthesis.

1.4 Mitochondrial Membranes

The number of mitochondria in various tissues reflects the physiological function of the tissue and determines its capacity, to perform aerobic metabolic functions.

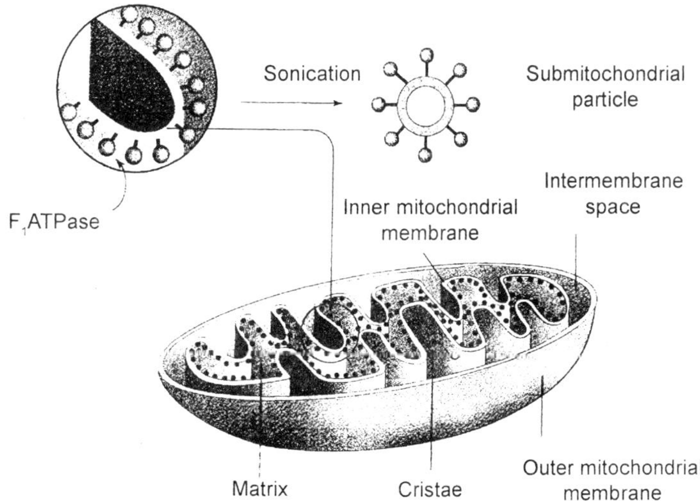

Diagram of various submitochondrial compartments.

For example, the erythrocyte has no mitochondria and does not possess the capacity to generate energy using oxygen as a terminal electron acceptor. On the other hand, cardiac tissue is a highly aerobic tissue, and it has been estimated that about one-half of the cytoplasmic volume of cardiac cells is composed of mitochondria. The liver is also highly dependent on aerobic metabolic processes for its various functions, and it has been estimated that mammalian hepatocytes contain between 800 and 2000 mitochondria. Mitochondria exist in a variety of different shapes, depending on the cell type from which they are derived. Mitochondria from liver are nearly spherical in shape, whereas those found in cardiac muscle are oblong or cylindrical.

Mitochondria are composed of two membranes, an outer and a highly invaginated inner membrane. The outer membrane is considered a rather simple membrane, composed of about 50% lipid and 50% protein, with relatively few enzymatic or transport functions.

The inner membrane is structurally and functionally much more complex than the outer membrane. Roughly 80% of the inner membrane is protein. It contains most of the enzymes involved in **electron transport** and oxidative phosphorylation, various dehydrogenases and several transport systems, which are involved in transferring substrates, metabolic intermediates and adenine nucleotides between the cytosol and the mitochondrial matrix.

Some enzymatic components are loosely associated with the inner membrane, whereas others are either tightly bound or are actual structural elements of the membrane. Hence there is a wide variability in the extent to which physical (ultrasonic irradiation or freezing and thawing), chemical (organic solvent or detergent treatment), or enzymatic (protease or lipase) treatments remove, release, or inactivate the enzymes associated with the inner membrane. Whereas the outer membrane presents little or no permeability barrier to substrate or nucleotide molecules of interest in energy metabolism, the inner membrane limits the types of substrates, intermediates, and nucleotides that can diffuse from the cytosol into the matrix compartment.

Various transport systems have been described in mitochondria (Figure below), some of which have been thoroughly characterized. The primary function of these transport systems is to facilitate the selective movement of various substrates and intermediates back and forth across the inner mitochondrial membrane from the cytosol to the mitochondrial matrix. Through the action of these transporters, various substrates and other molecules can be accumulated in the mitochondrial matrix since the transporters can facilitate the movement of the substrate against a concentration gradient. The importance of a mitochondrial transporter derives from involvement of the substance transported in a variety of mitochondrial metabolic processes.

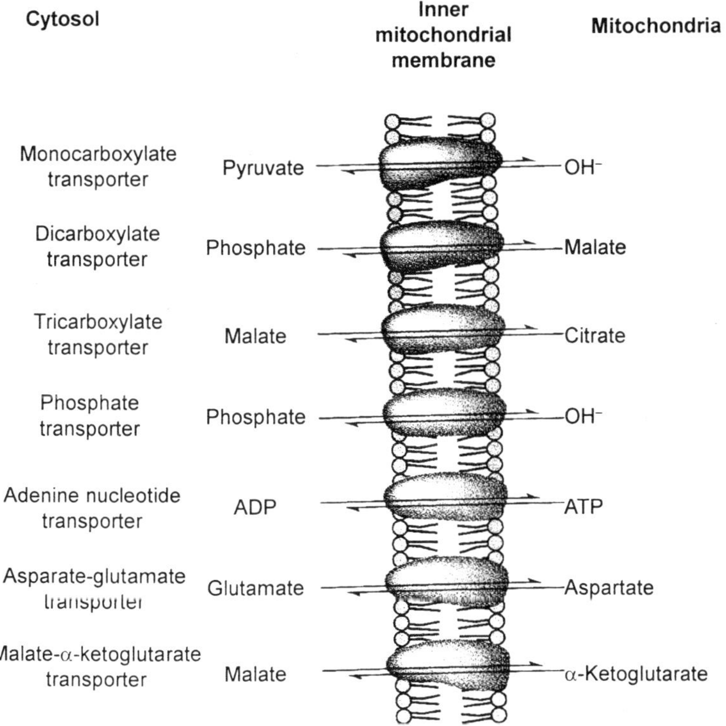

Mitochondrial metabolite transporters.

Mitochondria from most tissues possess a transport system for translocating Ca^{2+} across the mitochondrial inner membrane. It is difficult to overestimate the importance of the distribution/ redistribution of cellular calcium pools in different cell functions, such as muscle contraction, neural transmission, secretion, and hormone action. Calcium exists in distinct pools in the cell. The cytosol, mitochondria, endoplasmic reticulum, nuclei, and Golgi complex have their own pools of calcium. Some of the intracellular calcium is bound to nucleotides, metabolites, or membrane ligands, while a portion of the intracellular calcium is free in solution. A gradient of Ca^{2+} exists from outside to inside a cell. Estimates of intracellular cytosolic calcium are in the range of 10^{-7} M, whereas extracellular calcium is at least four orders of magnitude greater. Total intramitochondrial calcium has been estimated to be $\sim 10^{-4}$ M but the free ionic calcium concentration in the mitochondrion is in the range of 10^{-7} M. Hence processes involved in the alternate sequestering and release of an intracellular store of calcium can greatly influence intracellular calcium pools and various cell functions. Mitochondria accumulate rather large quantities of calcium at the expense of ATP hydrolysis, respiration, or the electrochemical gradient created across the mitochondrial membrane. Mitochondrial calcium transport is inhibited by low concentrations of lanthanides (trivalent metal cations) and by ruthenium red. Mg^{2+} competes with Ca^{2+} for the carrier in certain types of mitochondria. The current view is that there is a specific carrier in the inner mitochondrial membrane, which is likely a glycoprotein. The mitochondrial calcium carrier exhibits saturation kinetics, has a high affinity for calcium, and is highly specific for calcium. Permeant counterions such as phosphate or acetate stimulate calcium transport and allow the cation to be retained in the matrix. The most probable utility of the ability of mitochondria to accumulate calcium occurs during cellular injury when extracellular calcium enters the cell. Mitochondria can sequester the calcium to minimize the change in the cytosolic calcium level. Certain hormones may affect intracellular calcium distribution (e.g., epinephrine or vasopressin) as part of the mechanism of the hormone response; it is unlikely that the mitochondrial calcium pool contributes to the hormone-sensitive pool of calcium.

In summary, the inner mitochondrial membrane possesses a variety of transport systems involved in the movement of nucleotides, substrates, metabolites, and metal cations into and out of the mitochondrial matrix. These transport functions are essential for the complex cellular metabolic pathways and their regulation.

1.5 Cytochromes

Organisms that require oxygen (i.e., aerobic organisms) in their energy-generating functions possess various cytochromes that are involved in electron-transfer systems. Cytochromes are a class of proteins characterized by the presence of an iron-containing heme group bound to the protein. Unlike the heme group in hemoglobin or myoglobin in which the heme iron remains in the Fe^{2+} state, the iron in the heme of a cytochrome is alternately oxidized (Fe^{3+}) or reduced (Fe^{2+}) as it functions in the electron transport chain.

Cytochromes of mammalian mitochondria are designated as *a, b,* and *c* on the basis of the α band of their absorption spectrum and the type of heme group. Cytochrome *c* is a all protein (104

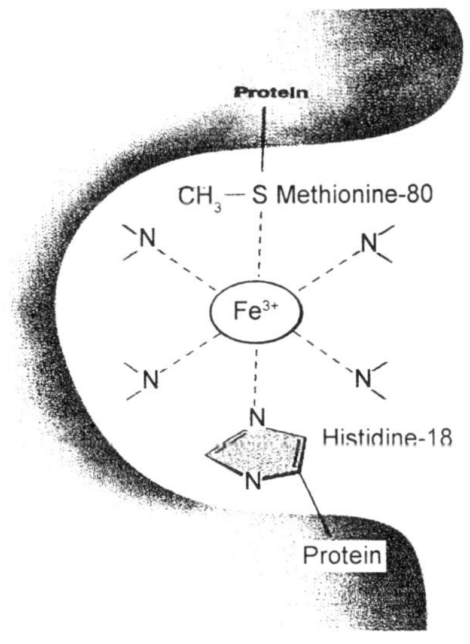

The six coordination positions of cytochrome *c*.

amino acid residues) with mol wt = 13,000. Amino acid sequences of cytochrome *c* from a great many species have been described and show that 20 out of 104 amino acid residues are invariant. The iron of the heme group in cytochrome *c* is coordinated between the four nitrogen atoms of the tetrapyrrole structure of the porphyrin group, whereas the fifth and sixth coordination positions are occupied by the methionine residue at position 80 and the histidine residue at position 18 of the protein.

Since all six coordination positions are filled in most of the cytochromes, binding of oxygen directly to the iron is prevented as is binding of respiratory inhibitors such as cyanide, azide, and carbon monoxide. The notable exception is cytochrome a_3, which is involved in the terminal step in mitochondrial electron transport. The heme group in cytochrome *c* is attached to the protein, not only by the fifth and sixth coordination positions of the heme iron, but also by the vinyl side chains of the protoporphyrin IX structure, from which hemes in cytochromes *a* and *c* are derived. These vinyl side chains are reduced by the addition of reduced sulfhydryls from cysteine residues at positions 14 and 17 in cytochrome *c* apoprotein. Hence the heme is covalently linked to the protein as well as being coordinated through the Fe^{2+} group in the heme.

The various electron-transferring proteins and other electron carriers that comprise the mitochondrial electron-transfer chain are arranged in a sequential pattern in the inner mitochondrial membrane. Reducing equivalents are extracted from substrates in the TCA cycle, the fatty acid β-oxidation sequence, and indirectly from glycolysis and passed sequentially through the electron transport chain to molecular oxygen. Electrons or reducing equivalents are fed into the electron transport chain at the level of NADH or coenzyme Q from the primary NAD^+- and FAD-linked dehydrogenase reactions and are transported to molecular oxygen through the cytochrome chain. This electron transport system is constructed so that the reduced member of one redox couple is oxidized by the oxidized member of the next component in the system.

$$NADH + H^+ + FMN \leftrightarrow FMNH_2 + NAD^+$$

or

Cytochrome *b* (Fe^{2+}) + cytochrome c_1 (Fe^{3+}) ↔
cytochrome *b* (Fe^{3+}) + cytochrome c_1 (Fe^{2+})

Note that electron transfer from NADH through coenzyme Q involves 2 e^-, whereas the reactions between coenzyme Q and oxygen involving the various cytochromes are 1 e^-, transfer reactions.

The component of the respiratory chain have characteristic absorption spectra that can be determined in suspensions of isolated mitochondria or submitochondiral particles using a dual-beam spectrophotometer. One of the light beams of the spectrophotometer was passed through a suspension of liver mitochondria, which was maintained under fully reduced conditions (e.g., substrate plus no oxygen), and the other beam was passed through an identical suspension in the presence of oxygen. The resulting spectrum is a difference spectrum of the reduced minus the oxidized states of the mitochondrial respiratory chain.

During transfer of electrons from the NADH–NAD^+ couple ($E_0' = -0.32$) to molecular oxygen ($E_0' = +0.82$) there occurs an oxidation–reduction potential decrease of 1.14 V. This drop in potential occurs in discrete steps as reducing equivalents or electrons are passed between the different segments of the chain. There is at least a 0.3-V decrease in potential between each of the three coupling or phosphorylation sites. A potential drop of 0.3 V is more than sufficient to accommodate synthesis of a high-energy phosphate bond of ATP. For example.

$$\Delta E_0' = 0.3 \text{ V}$$

$$\Delta G^\circ = -nF\Delta E_0'$$

$$\Delta G^\circ = 2 \times 23.062 \times 0.3$$

$$\Delta G^\circ = -13.8 \text{ kcal mol}^{-1}$$

Various components of the electron transport chain are located asymmetrically in the mitochondrial membrane.

1.6 Tricarboxylic Acid Cycle

The initial step of the cycle is catalyzed by **citrate synthase.** This is a highly exergonic reaction and commits acetyl groups to **citrate** formation and complete oxidation in the Krebs cycle. As shown below citrate synthase involves condensation of an acetyl moiety and the α-keto function of the dicarboxylic acid **oxalo-acetate.** Citrate synthase (mol wt 100,000) is in the mitochondrial matrix.

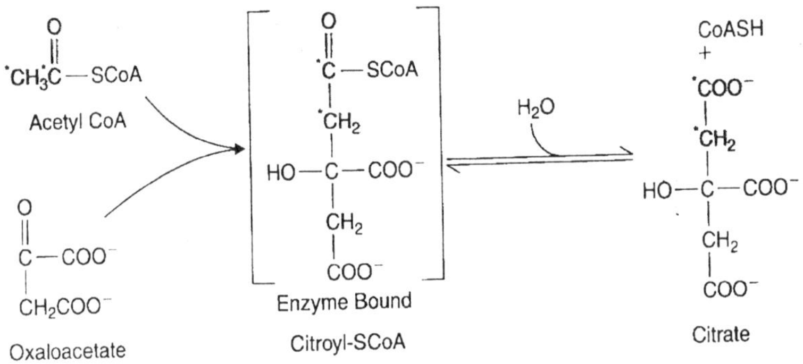

$\cdot CH_3\overset{\cdot}{C}(=O)—SCoA$ (Acetyl CoA)

$O=C(—COO^-)—CH_2COO^-$ (Oxaloacetate)

[$\cdot C(=O)—SCoA$ / $\cdot CH_2$ / $HO—C—COO^-$ / CH_2 / COO^-] Enzyme Bound Citroyl-SCoA

H_2O

CoASH + $\cdot COO^-$ / $\cdot CH_2$ / $HO—C—COO^-$ / CH_2 / COO^- Citrate

Citrate Synthase.

The equilibrium of this reaction is far toward citrate formation with a ΔG° near -9 kcal mol^{-1}. The citroyl-SCoA intermediate is not released from the enzyme during the reaction and remains bound to the catalytic site on citrate synthase. The citrate synthase reaction is considerably displaced from equilibrium under *in situ* conditions, which makes this step a primary candidate for regulatory modulation. The purified enzyme is regulated (inhibited) by ATP, NADH, succinyl CoA, and long-chain acyl CoA derivatives, but these effect have not been demonstrated in intact metabolic systems under physiological conditions.

The most probable means for regulating the citrate synthase reaction is availability of its two substrates, acetyl CoA and oxaloacetate. Note the many important fates and effects of citrate in energy and biosynthetic metabolism, citrate is a regulatory effector of other metabolic pathways and a source of carbon and reducing equivalents for various synthetic purposes.

Citrate synthase reacts with **monofluoroacetyl CoA** to form **monofluorocitrate,** a potent inhibitor of the next step in the cycle, the **aconitase** reaction. In fact, whether monofluorocitrate is synthesized *in situ* as a result of **fluoroacetate** poisoning or administered experimentally, a nearly complete block of TCA cycle activity is observed.

Citrate is converted to isocitrate in the aconitase reaction:

COO^- — CH_2 — HO—C—COO^- — CH_2 — COO^- (Citrate) $\underset{H_2O}{\rightleftharpoons}$ [COO^- — CH_2 — C—COO^- = HC — COO^-] (*cis*-Aconitate) $\underset{H_2O}{\rightleftharpoons}$ COO^- — CH_2 — H—C—COO^- — HO—C—H — COO^- (Isocitrate)

Aconitase.

This reaction involves generation of an enzyme-bound intermediate, ***cis*-aconitate.** At equilibrium there exist 90% citrate, 3% *cis*-aconitate, and 7% isocitrate; hence the equilibrium of aconitase lies toward citrate formation. Although the aconitase reaction does not require cofactors, it requires ferrous (Fe^{2+}) iron in its catalytic mechanism. This Fe^{2+} is involved in an **iron-sulfur center,** which is an essential component in the hydratase activity of aconitase.

Isocitrate dehydrogenase catalyzes the first dehydrogenase reaction in the TCA cycle. Isocitrate is converted to **α-keto-glutarate** in an oxidative decarboxylation reaction. In this step of the cycle the initial (of two) CO_2 is produced and the initial (of three) NADH + H^+ are generated.

2

Composition and Structure of Proteins

2.1 Introduction

The Duch chemist, **Gerardus Mulder,** is accredited with the adoption of the word **protein,** in 1838. Mulder had been studying the properties of a substance from milk and eggs which coagulated upon heating. This substance was common to animal tissues and plant juices. Mulder considered the substance to be of primary importance to all living organisms. The name protein (Greek *proteios* = of first quality) was actually suggested to him by the Swedish chemist, Baron Jöns Jakob Berzelius, who had earlier discovered the element selenium (1817), silicon (1824) and thorium (1828).

Proteins are the most abundant macromolecules found within cells. A protein may be considered as a unique polymer of amino acids. The different constituent amino acids determine the overall chemical and structural properties of a protein. There are over 300 catalogued amino acids but only 20 (plus a few derivatives) occur in proteins. Some proteins may contain derivatives of common amino acids, e.g. collagen contains 5-hydroxylysine and hydroxyprolines. Such derivatives are produced in reactions located in the endoplasmic reticulum during or subsequent to ribosomal protein synthesis. These reactions are termed **co- or post-translational modifications.**

Proteins perform a surprising variety of essential functions in mammalian organisma. These may be grouped into dynamic and structural. Dynamic functions include transport, metabolic control, contraction, and catalysis of chemical transformations. In their structural functions, proteins provide the matrix for bone and connective tissue, giving structure and form to the human organism.

An important class of dynamic proteins are the enzymes. They catalyze chemical reactions, converting a substrate to a product at the enzyme's active site. Almost all of the thousands of chemical reactions that occur in living organisms require a specific enzyme catalyst to ensure that reactions occur at a rate compatible with life. The character of any cell is based on its particular chemistry, which is determined by its specific enzyme composition. Genetic traits are expressed through synthesis of enzymes, which catalyze reactions that establish the phenotype. Many genetic disease result from altered levels of enzyme production or specific alterations to their amino acid sequence. Transport is another major function for proteins. Particular examples discussed in greater detail in this text are hemoglobin and myoglobin, which transport oxygen in blood and in muscle, respectively. Transferrin transports iron in blood. Transport proteins bind and carry steroid hormones in blood from their site of synthesis to their site of action. Many drugs and toxic compounds are transported bound to proteins. Proteins participate in contractile mechanisms. Myosin and actin function in muscle contraction.

Proteins have a protective role through a combination of dynamic functions. Immunoglobulins and interferon are proteins that protect the human against bacterial or viral infection. Fibrin stops the loss of blood on injury to the vascular system.

Many hormones are proteins or peptides. Protein hormones include insulin, thyrotropin, somatotropin (growth hormone), luteinizing hormone, and follicle-stimulating hormone. Many diverse polypeptide hormones have a low molecular weight (<5000) and are referred to as peptides. In general, the term **protein** is used for molecules composed of over 50 amino acids and the term **peptide** is used for molecules of less than 50 amino acids. Important peptide hormones include adrenocorticotropin hormone, antidiuretic hormone, glucagon, and calcitonin.

Proteins control and regulate gene transcription and translation. These include histones that are closely associated with

DNA, repressor and enhancer transcription factors that control gene transcription, and proteins that form a part of the heteronuclear RNA particles and ribosomes.

Structural proteins function in "brick-and-mortar" roles. They include collagen and elastin, which form the matrix of bone and ligaments and provide structural strength and elasticity to organs and the vascular system. α-Keratin forms the structure of epidermal tissue.

An understanding of both the normal functioning and the pathology of the mammalian organism requires a clear understanding of the properties of the proteins.

2.2 Amino Acids

The **first amino acid** to be discovered was asparagine from asparagus tissue in 1806. The first amino acid to be identified from protein was leucine by Proust in 1819. Cystine (two cysteine molecules linked by a disulphide bond) was extracted from urinary calculi in 1810 but it was not obtained from protein until 1899. The last amino acid to be discovered in protein was threonine in 1936. However, the proline derivative, 3-hydroxy-proline, present in collagen was not identified until 1962.

The naming of amino acids has not been systematic. Glycine was thus named because it tastes sweet (Greek *glycos* = sweet). Tyrosine was first isolated from cheese (Greek *tyros* = cheese) and glutamic acid was found in wheat gluten.

Amino acids are the **building blocks** of proteins. Some amino acids which do not occur in proteins may function in cellular metabolism, e.g. citrulline and ornithine.

Amino Acids in Proteins

Proline, an imino acid, is usually included along with the amino acids because of its occurrence in proteins. The **amino acids found in proteins** may be referred to by the use of one-letter or three-letter codes.

Abbreviations for the Amino Acids

Amino Acid	*Abbreviation*	
	Three Letter	*One Letter*
Alanine	Ala	A
Arginine	Arg	R
Asparagine	Asn	N
Aspartic	Asp	D
Asparagine or aspartic	Asx	B
Cysteine	Cys	C
Glycine	Gly	G
Glutamine	Gln	Q
Glutamic	Glu	E
Glutamine or glutamic	Glx	Z
Histidine	His	H
Isoleucine	Ile	I
Leucine	Leu	L
Lysine	Lys	K
Methionine	Met	M
Phenylalanine	Phe	F
Proline	Pro	P
Serine	Ser	S
Threonine	Thr	T
Tryptophan	Trp	W
Tyrosine	Tyr	Y
Valine	Val	V

It is notable that all the different types of proteins are initially synthesized as polymers of only 20 amino acids. These **common amino acids** are defined as those for which at least one specific codon exists in the DNA genetic code. There are 20 amino acids for which DNA codons are known. Transcription and translation of the DNA code result in polymerization of amino acids into a specific linear sequence` characteristic of a protein. In addition to the common amino acids, proteins may contain **derived amino acids,** which are usually formed by an enzyme-facilitated reaction on a common amino acid after that amino acid has been incorporated into a protein structure. Examples of derived amino acids are cystine, desmosine and isodesmosine found in elastin, hydroxyproline and hydroxylysine found in collagen, and γ-carboxyglutamate found in prothrombin.

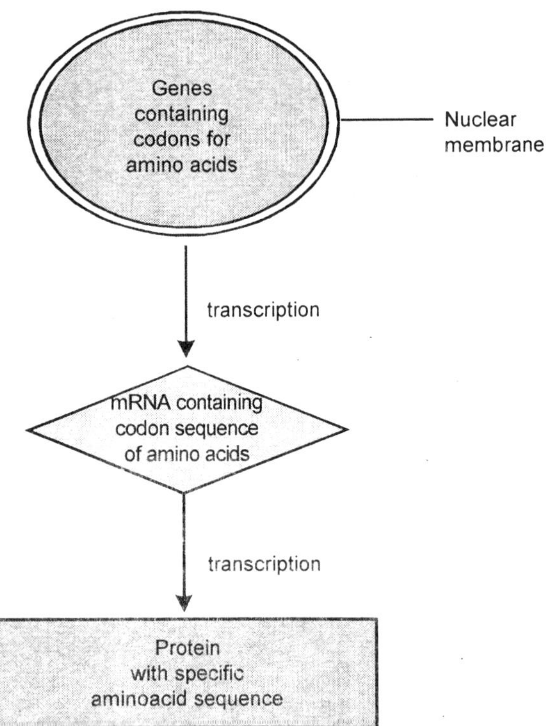

Genetic information is transcribed from a DNA sequence into mRNA and then translated to the amino acid sequence of a protein.

Common amino acids, contain in common a central ***alpha*** **(α)-carbon** atom to which a carboxylic acid group, an amino group, and a hydrogen atom are covalently bonded. In addition, the α-carbon atom is bound to a specific chemical group, designated R and called the side chain, that uniquely defines each of the 20 common amino acids. The α-amino group is protonated and in its ammonium ion form; the carboxylic acid group is in its unprotonated or carboxylate ion form

Alkyl amino acids have alkyl group side chains and include glycine, alanine, valine, leucine, and isoleucine. **Glycine** has the simplest structure, with R = H. **Alanine** contains a methyl (CH_3—) side chain group. **Valine** has an isopropyl R group. The leucine and isoleucine R groups are butyl groups that are structural isomers of each other. In **leucine** the branching in the isobutyl side chain occurs on the *gamma* (γ)-carbon of the amino acid. In isoleucine it is branched at the *beta* (β)-carbon.

The aromatic amino acids are phenylalanine, tyrosine, and tryptophan. The **phenylalanine** R group contains a benzene ring, **tyrosine** contains a phenol group, and the **tryptophan** R group contains the heterocyclic structure, indole.

2.3 Characteristic of Amino Acids

The common amino acids with the general structure have four substituents (R, H, COO^-, NH_3^+) covalently bonded to the α-carbon atom in the α-amino acid structure. A carbon atom with four different substituents arranged in a tetrahedral configuration is asymmetric and exists in two enantiomeric forms. Thus each of the amino acids exhibits optical isomerism except glycine, in which R=H and thus two of the four substituents on the α-carbon atom are hydrogen.

Using the Fischer projection to show the direction in space of the tetrahedrally arranged α-carbon substituents. The α-COO^- group is directed up and behind the plane of the page, and the R group is directed down and behind the plane of the page. The α-H and α-NH_3^- groups are directed toward the reader. An amino acid held in this way projects its α-NH_3^+ group either to the left or right of the α-carbon atom. By convention, if the α-NH_3^+ is projected to the left, the amino acid has an L absolute configuration. Its optical enantiomer, with α-NH_3^+ projected toward the right, has a D absolute configuration. In mammalian proteins only amino acids of L configuration are found. The L and D designations refer to the ability to rotate polarized light to the left (L, levo) or right (D, dextro) from its plane of polarization. As the amino acids in proteins are asymmetric, the proteins that contain them also exhibit asymmetric properties.

Polymerization of the 20 common amino acids into polypeptide chains in cells is catalyzed by enzymes and is associated with the ribosomes. Chemically, this polymerization is a dehydration reaction. The α-carboxyl group of an amino acid with side chain R_1 forms a covalent **peptide bond** with the α-amino group of the amino acid with side chain R_2 by elimination of a molecule of water. The **dipeptide** (two amino acid residues joined by a single peptide bond) can then form a second peptide bond through its terminal carboxylic acid group and the α-amino of a third amino acid (R_3), to generate a tripeptide. Repetition of this process generates a **polypeptide** or protein of specific amino acid sequence (R_1-R_2-R_3-R_4-. . . .R_n). The amino acid sequence of the polypeptide chains is the **primary structure** of the protein, and it is predetermined by the DNA sequence of its gene. It is the unique primary structure that enables a polypeptide chain to fold into a specific three-dimensional structure that gives the protein its chemical and physiological properties.

A peptide bond can be represented using two **resonance isomers**. In Structure I, a double bond is located between the carbonyl carbon and carbonyl oxygen (C′=O), and the carbonyl carbon to nitrogen (C′—N) linkage is a single bond. In Structure II, the carbonyl carbon to oxygen bond (C′—O^-) is a single bond and the bond located between the carbonyl carbon and nitrogen is a double bond (C′=N). In Structure II there is a negative charge on the oxygen and a positive charge on the nitrogen. Actual peptide bonds are a resonance hybrid of these two electron isomer structures, the carbonyl carbon to nitrogen bond having a 50% double-bond character. The hybrid bond is supported by spectroscopic measurements and X-ray diffraction studies, the latter showing that the carbonyl carbon to nitrogen peptide bond length (1.33 Å) is approximately half-way between that found for a C—N single bond (~1.45 Å) and a C=N double bond (~1.25 Å).

A consequence of this partial double-bond character is that, as for normal double-bond structures, rotation does not occur about the carbonyl carbon to nitrogen of a peptide bond at physiological

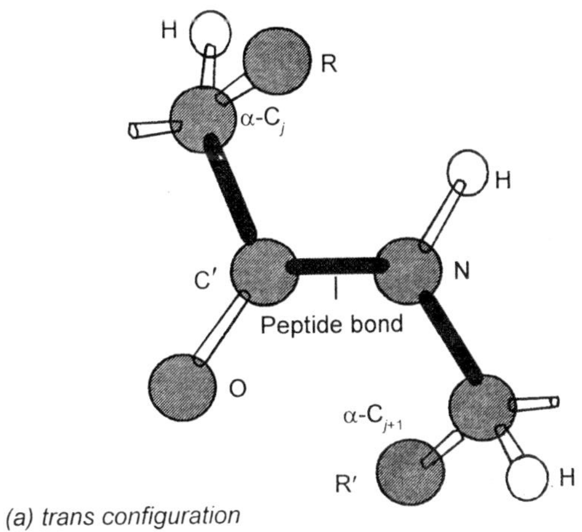

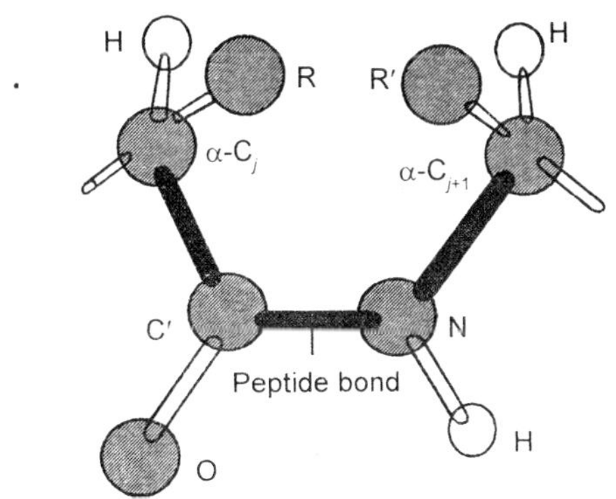

(*a*) *Trans*-peptide bond and (*b*) the rare *cis*-peptide bond.

temperatures. Also, a consequence of the C′=N double-bond's chemistry is that the atoms attached to C′ and N all lie in a common plane. Thus a polypeptide chain is a polymer of peptide bond planes

$$\mathrm{-C-C'(O^{-})=N^{+}(H)-C-}$$

interconnected at the α-carbon atoms. The α-carbon interconnects peptide bonds through single bonds that allow rotation of adjacent peptide planes with respect to each other. Each **amino acid residue** contributes one α-carbon (two single bonds and a peptide bond), to the polypeptide chain. The term residue refers to the atoms contributed by an amino acid to a polypeptide chain including the atoms of the side chain.

The peptide bond shows a ***trans* configuration** between the oxygen (O) and the hydrogen (H) atoms of the peptide bond. This is the most stable configuration for the peptide bond with the two side chains (R and R′) also in *trans*. The ***cis* configuration** brings the two side chain groups to the same side of the C′=N bond, where unfavourable repulsive steric forces occur between the two side chain (R) groups. Accordingly, *trans*-peptide bonds are always found in proteins except where there are proline residues. In proline the side chain is linked to its α-amino group, and the *cis*- and *trans*-peptide bonds with the proline α-imino group have near equal energies. The configuration of the peptide bond actually found for a proline in a protein will depend on the specific forces generated by the unique folded three-dimensional structure of the protein molecule.

One of the largest natural polypeptide chains in humans is that of apolipo-protein B-100, which contains 4536 amino acid residues in one polypeptide chain. Chain length alone, however, does not determine the function of a polypeptide. Many small peptides with less than ten amino acids perform important biochemical and physiological functions in humans. Primary structures are written in a standard convention and sequentially numbered from their NH_2-terminal end toward their COOH-terminal end, consistent with the order of addition of the amino acid to the chain during biosynthesis. Accordingly, for thyrotropin-releasing hormone the glutamic acid residue written on the left is the NH_2-terminal amino acid of the tripeptide and is designated amino acid residue 1 in the sequence. The proline is the COOH-terminal amino acid and is designated residue 3. The defined direction of the polypeptide chain is from Glu to Pro (NH_2-terminal amino acid to COOH-terminal amino acid).

The **ionization characteristics** of the amino acids in aqueous media are denoted by their **pK_a values**. Observe the similarity in values for pK_{a1} of the α-carboxyl group and the pK_a of the α-amino group of different amino acids. Amino acids respond to changes in their aqueous environment in accordance with the **acid-base principles**. In aqueous solutions at pH 1, the amino acid is fully protonated. Titration of this amino acid by the addition of one equivalent of equimolar base will release a proton from the carboxylic acid group. The addition of a second equivalent of equimolar base deprotonates the amino group. For example, alanine at low pH is fully protonated and carries a net charge of +1. Increasing the pH by the addition of one equivalent of base results in the complete ionization of the carboxyl group and the entire molecule exhibits no net charge. The addition of a second equivalent of base causes the removal of the proton from the NH_3^+ group of alanine which now has a net charge of –1. Thus the amino acids display amphoteric behaviour, i.e. they are capable of acting as an acid or base. The parameters pK_{a1} and pK_{a2} can be determined at the appropriate half-equivalence point.

With **neutral and monoamino-dicarboxylic acids** at the first equivalence point, the positive charge balances the negative charge. The molecules are therefore in their dipolar ionic state and are called **zwitterions** (German for hybrid ion). Since their net charge is 0, the zwitterions cannot migrate in an electric field, e.g. during the technique of electrophoresis. This pH at which a molecule exhibits no net charge is termed its **isoelectric point (pI).**

The **titrations of amino acids with dissociable R groups** conform to the same principles but require three equivalents of bases. The pI of any amino acid can be calculated according to the following formulae:

- *amino acids* other than basic amino acids: pI = 0.5 (pK_{a1} + pK_{a2});
- *basic amino acids:* pI = 0.5 (pK_{a2} + pK_{a3}).

2.4 Structure of Proteins

The **primary structure** (amino acid sequence) of a protein is required for an understanding of a protein's structure, its mechanism of action at a molecular level, and its relationship to other proteins with similar physiological roles. The primary structure of **insulin** illustrates the value of this knowledge for understanding a protein's biosynthesis and physiological forms. Insulin is produced in pancreatic islet cells as a single chain precursor, **proinsulin**. The polypeptide chain contains 86 amino acids and 3 intrachain cystine disulfide bonds. It is transformed into biologically active insulin by proteolytic modifications in its primary structure as it is secreted from the islet cells. Proinsulin is cleaved by proteases present in the islet cells that cleave two peptide bonds in proinsulin between residues 30 and 31 and 65 and 66. This releases a 35 amino acid segment (the **C-peptide**) peptide is further processed in the pancreatic islet cells by proteases that hydrolyze a dipeptide from the COOH terminal and a second dipeptide from the NH_2 terminal of the C-peptide. The modified C-peptide is secreted into the blood with the active insulin. Besides giving information on the pathway for formation of active insulin, knowledge of primary structures shows the role of particular amino acids in the structure of insulin through comparison of the primary structures of the insulins from different animal species. The aligned primary structures show a residue identity in most amino acid positions, except for residues 8, 9, and 10 of the A chain and residue 30 of the B chain. Amino acids in these positions vary widely in different animal insulins and apparently do not affect the biological properties of the insulin molecule. Other amino acids of the primary structure are rarely substituted, suggesting that they have an essential role in insulin function.

Comparison of primary structures is commonly used to predict the similarity in structure and function between proteins. Sequence comparisons typically require aligning sequences to maximize the number of identical residues while minimizing the number of insertions or deletions required to achieve this alignment. Two sequences are termed **homologous** when their sequences are highly alignable. In its correct usage homology only refers to proteins that have evolved from the same gene. **Analogy** is used to describe sequences from proteins that are structurally similar but for which no evolutionary relationship has been demonstrated.

Secondary structure refers to the local three-dimensional folding of the polypeptide chain in the protein. The polypeptide chain in this context is the covalently interconnected atoms of the peptide bonds and α-carbon linkages that sequentially link the amino acid residues of the protein. Side chains are not considered at the level of secondary structure. **Tertiary structure** refers to the three-dimensional structure of the polypeptide. It includes the conformational relationships in space of the side chains and the

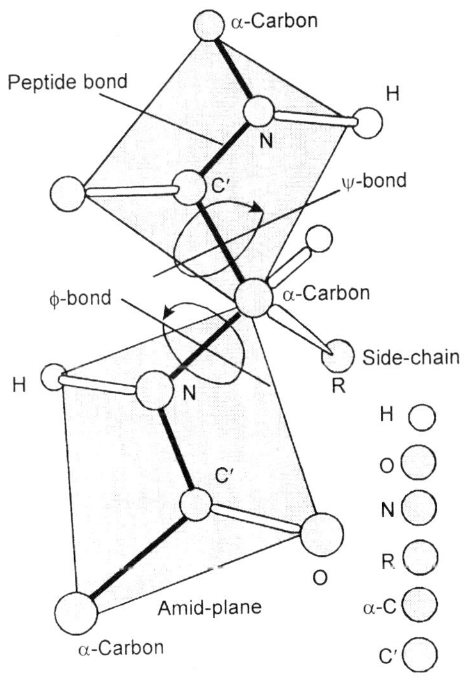

Polypeptide chain showing ϕ, ψ, and peptide bonds for residue R_i within a polypeptide chain.

geometric relationship between distant regions of the polypeptide chain. **Quaternary structure** refers to the structure and interactions of the noncovalent association of discrete polypeptide subunits into a multisubunit protein. Not all proteins have a quaternary structure.

Proteins generally assume unique secondary, tertiary, and quaternary conformations as determined by their particular amino acid sequence and termed the **native conformation.** Folding of the primary structure into the native conformation occurs, in most cases, spontaneously through noncovalent interactions. This unique conformation is the one of lowest total Gibbs free energy kinetically accessible to the polypeptide chain(s) for the particular conditions of ionic strength, pH, and temperature of the solvent in which the folding occurs. Chaperone proteins may facilitate the rate of protein folding.

The conformation of a polypeptide chain may be described by the rotational angles of the covalent bonds that contribute to the polypeptide chain. These are the bonds contributed by each of the amino acids between (1) the nitrogen and α-carbon and (2) the α-carbon and the carbonyl carbon. The first of these is designated the **phi (ϕ) bond** and the second is called the **psi (ψ) bond** for an amino acid residue in a polypeptide chain. The third bond contributed by each amino acid to the polypeptide chain is the peptide bond. As previously discussed, due to the partial double-bond character of the C′$\overset{\cdots}{-}$N bonds, there is a barrier to free rotation about this peptide bond.

Regular secondary structure conformations in segments of a polypeptide chain occur when all ϕ bond angles in that polypeptide segment are equal, and all the ψ bond angles are equal.

The α-helix and β-structure conformations for polypeptide chains are the most thermodynamically stable of the regular secondary structures. However, a particular sequence may form regular conformations other than α-helical or β-structure. There are also regions of unordered secondary structure, in which neither the ϕ bond angles nor the ψ bond angles are equal. Proline interrupts α-helical conformations since the pyrrolidine side chain of proline sterically interacts with the amino acid preceding it in the polypeptide sequence when in an α-helical structure. This repulsive steric interaction tends to prevent formation of α-helical structure in sections of a polypeptide chain that contain proline.

Helical structures of polypeptide chains are characterized by the number of amino acid residues per turn of helix (n) and the distance between α-carbon atoms of adjacent amino acids measured parallel to the axis of the helix (d). The **helix pitch** (p), defined as the product of $n \times d$, then measures the distance between repeating turns of the helix on a line drawn paralled to the helix axis.

$$p = n \times d$$

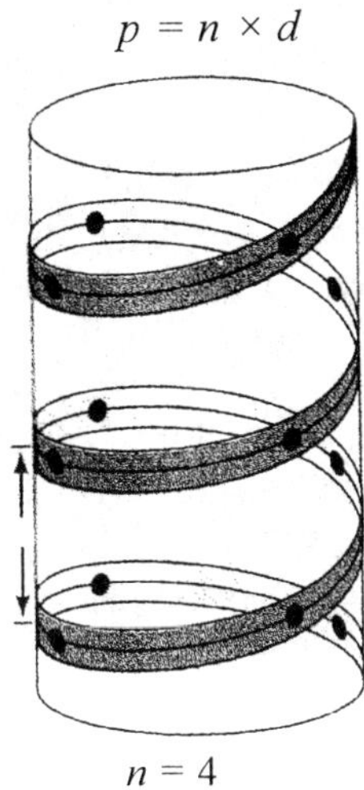

$n = 4$

The helix pitch (p) for a helix with n = 4.

The **tertiary structure** of a protein is the three-dimensional structure of a protein. It includes the geometric relationship between distant segments of primary structure and the relationship of the side chains with one another in three-dimensional space. As an example of a protein's tertiary structure, the structure for trypsin. The ribbon structure shows the conformation of polypeptide strands and the overall pattern of polypeptide chain folding (supersecondary structure). The tertiary structure is then further built upon by showing the side chain groups and their interconnections with a stick model. Active site catalytic side chains are shown, which include the hydroxymethyl group of serine, the imidazole of histidine, and the carboxylate containing side chain of aspartate. Although these catalytic residues are widely separated in the primary structure, the folded tertiay structure brings them together in space to form the catalytic site. A space-filling model shows C, N, and O atoms represented by balls of radius proportional to their van der Waals radius.

The tertiary structure of trypsin conforms to the general rules of folded proteins. Hydrophobic side chains are generally in the

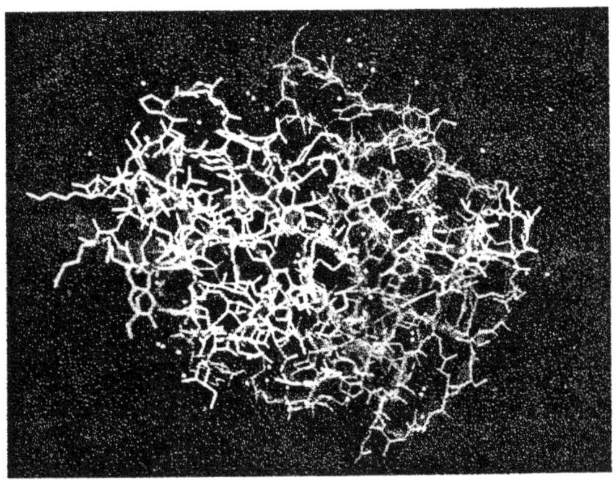

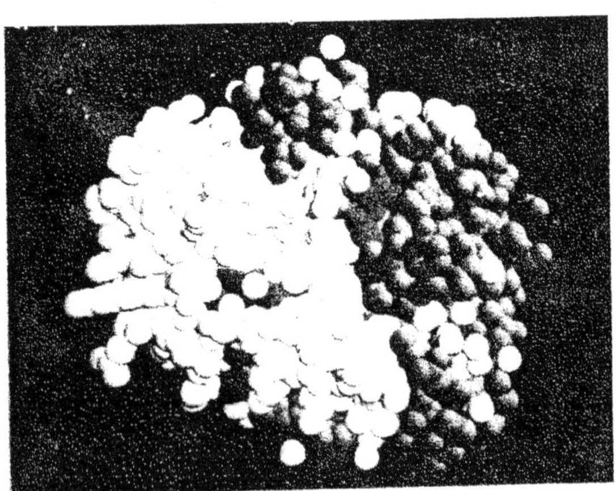

interior of the structure, away from the water interface. Ionized side chains occur on the outside of a protein structure, where they are stabilized by water of solvation. Within the protein structure (not shown) are buried water molecules, noncovalently associated, which exhibit specific arrangements. A large number of water molecules form a solvation shell around the outside of the protein.

A long polypeptide strand often folds into multiple compact semi-independent folded regions or **domains**, each domain having a characteristic compact geometry with a hydrophobic core and polar outside. They typically contain 100-150 contiguous amino acids. The domains of a **multidomain protein** may be connected by a segment of the polypeptide chain lacking regular secondary structure. Alternatively, the dense spherical folded regions are separated by a cleft or less dense region of tertiary structure. There are two folded domains in the trypsin molecule with a cleft between the domains that includes the substrate-binding catalytic site of the protein. An active site within an interdomain interface is an attribute of many enzymes. Different domains within a protein can move with respect to each other. Hexokinase, which catalyzes phosphorylation of a glucose molecule by adenosine triphosphate (ATP), has the glucose-binding site in a region between two domains. When the glucose binds in the active site, the surrounding domains move to enclose the substrate to trap it for phosphorylation. In enzymes with more than one substrate or allosteric effector site, the different sites may be located within different domains. In multifunctional proteins, each domain performs a different task.

A protein can adopt a range of conformations for a particular amino acid sequence. Although each native structure is unique, a comparison of the tertiary structures of different proteins solved by X-ray crystallography shows similar arrangements of secondary structure motifs that form the tertiary structures of domains. Thus proteins unrelated by function, sequence, or evolution show similar patterns of arrangement of their secondary structures.

Quaternary structure refers to the arrangement of polypeptide chains in a multichain protein. The subunits in a quaternary structure must be in noncovalent association. α-Chymotrypsin contains three polypeptide chains covalently joined together by interchain disulfide bonds into a single covalent unit and therefore does not have a quaternary structure. Myoglobin consists of a single polypeptide chian and has no quaternary structure. However, hemoglobin A contains four polypeptide chains ($\alpha_2\beta_2$) held together noncovalently in a specific conformation as required for its function. Thus hemoglobin has a quaternary structure. Aspartate transcarbamylase has a quaternary structure comprised of 12 polypeptide subunits. The poliovirus protein coat contains 60 polypeptide subunits, and the tobacco mosaic virus protein has 2120 polypeptide subunits held together noncovalently in a specific structural arrangement.

2.5 Types of Proteins

Now let us study about the fibrous proteins, which are nonglobular and have a low water solubility. Fibrous proteins include collagen Elastin, α-keratin and Tropomyosin.

Globular proteins have a spheroidal shape, variable molecular weights and relatively high water solubility.

Collagen

Collagen is one of the most abundant fibrous proteins in the vertebrate body. The importance of collagen may also be demonstrated by the consequences of over 400 mutations identified in only six types of collagen. There are at least 19 genetically distinct types of collagen, labelled Types I-XIX. The numbers reflect the chronological order of discovery. Collagen has a distinctive structure which is also found in a limited number of **'collagen-like' proteins,** e.g. component C1q of the complement system, mannan-binding protein, the pulmonary surfactant proteins (SP-A and SP-D) and the conglutinin tetramer. Early electron microscopy studies suggested that collagen had a repetitive uniform structure irrespective of its source. However, this is not the case but simply appeared so because the tissues examined such as adult skin, bone, tendon, placenta and cornea contain large amounts of Type I collagen.

The amino acid composition of collagen is quite different from that of a typical globular protein. The amino acid composition of type I skin collagen and of globular proteins ribonuclease and hemoglobin. Skin collagen is comparatively rich in glycine (33% of its amino acids), proline (13%), the derived amino acid 4-hydroxyproline (9%), and another derived amino acid 5-hydroxlysine (0.6%). Hydroxyproline is unique to collagens being formed enzymatically from prolines within a collagen polypeptide chain. The enzyme-catalyzed hydroxylation of proline requires the presence of ascorbic acid (vitamin C); thus in vitamin C deficiency (scurvy) there is poor synthesis of new collagen. Most hydroxyprolines in a collagen have the hydroxyl group in the 4-position (γ-carbon) of the proline structure, although a small amount of 3-hydroxyproline is also formed. Collagens are glycoproteins with carbohydrate covalently joined to the derived amino acid, 5-hydroxylysine, by an *O*-glycosidic bond through the δ-carbon hydroxyl group. Formation of 5-hydroxylysine from lysines and addition of the carbohydrate to the 5-hydroxylysine occur after polypeptide chain formation but prior to the folding of the collagen chains into their unique supercoiled structure.

Elastin

Elastin is a fibrous protein with Allysine-Generated Cross-links.

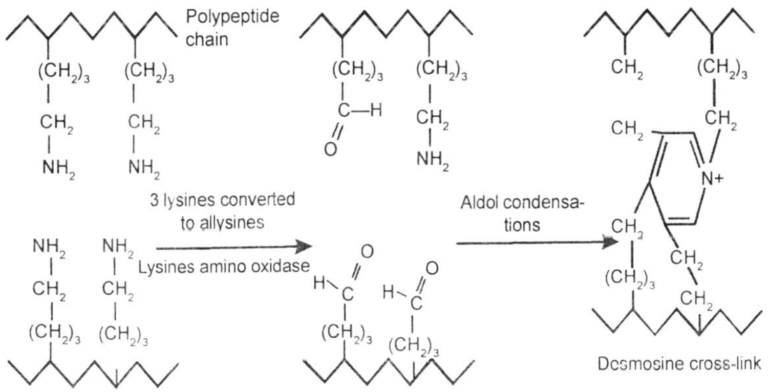

Polypeptide chain
3 lysines converted to allysines
Lysines amino oxidase
Aldol condensations
Desmosine cross-link

Formation of desmosine covalent cross-link in elastin from lysine and allysines.

Elastin gives tissues and organ the capacity to stretch without tearing. It is classified as a fibrous protein because of its structural function and relative insolubility. It is abundant in ligaments, lungs, walls of arteries, and skin. Elastin does not contain repeating sequences of Gly-Pro-Y or Gly-X-Hyp and does not fold into either a polyproline helix or a superhelix. It appears to lack a regular secondary structure, but rather contains an unordered coiled structure in which amino acid residues within the folded structure are highly mobile. The highly mobile, kinetically free, though extensively cross-linked structure gives the protein, a rubber like elasticity.

As in collagen, allysines form cross-links in **elastin**. An extracellular **lysine amino oxidase** converts lysine side chains of elastin to allysines. The amino oxidase is specific for lysines in the sequence -Lys-Ala-Ala-Lys- and –Lys-Ala-Ala-Ala-Lys-. Three allysines and an unmodified lysine in these sequences, from different regions in the polypeptide chains, react to form the heterocyclic structure of **desmosine** or **isodesmosine.** The desmosines covalently cross-link the polypeptide chains in elastin fibers.

Now let us deal with the other two types of fibrous proteins.

α-Keratin and Tropomyosin

α-Keratin and tropomyosin are fibrous proteins in which each chain has an α-helical conformation α-Keratin is found in the epidermal layer of skin, in nails, and in hair. Tropomyosin is a component of the thin filament in muscle tissue. Analysis of the α-helical sequences in both these proteins shows the tandem repetition of seven amino acid segments, in which the first and

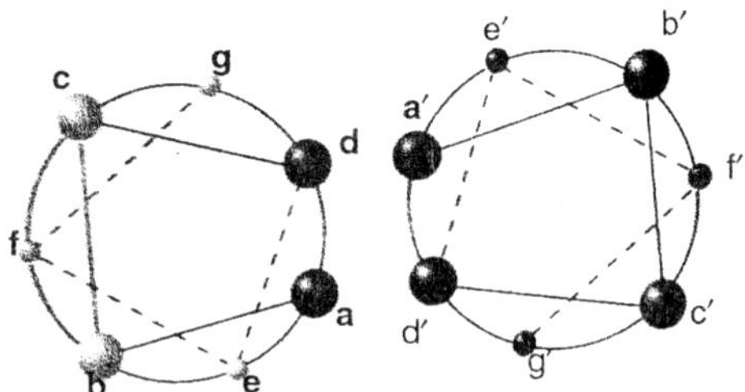

Interaction of an apolar edge of two chains in α-helical conformation as in α-keratin and tropomyosin

fourth amino acids have hydrophobic side chains and the fifth and seventh polar side chains. The reiteration of hydrophobic and polar side chains in seven amino acid segments is symbolically represented by the formulation (a-b-c-d-e-f-g)$_{t}$, where residues a and d are hydrophobic amino acids, and residues e and g are polar or ionized side chain groups. Since a seven amino acid segment represents two complete turns of an α-helix (n = 3.6), the apolar residues at a and d align to form an apolar edge along one side of the α-helix. This apolar edge in α-keratin interacts with polypeptide apolar edges of other α-keratin chains to form a superhelical structure containing two or three polypeptide chains. Each strand also contains a polar edge, due to residues e and g, that interacts with the water solvent on the outside of the superhelix and also stabilizes the superhelical structure. Similarly, two tropomyosin polypeptide strands in α-helical conformation wind around each other to form a tropomyosin superhelical structure.

Thus collagen, α-keratin, and tropomyosin molecules are multistrand structures in which polypeptide chains with a highly regular secondary structure (polyproline type II helix in collagen, α-helix in α-keratin and tropomyosin) are wound around each other to form a rod-shaped supercoiled conformation. In turn, the supercoiled molecules are aligned into multimolecular fibrils stabilized by covalent cross-links. The amino acid sequences of the chains are repetitive, generating edges on the cylindrical surfaces of each of the chains that stabilize a hydrophobic interaction between the chains required for generation of the supercoiled conformation.

2.6 The Folding of Proteins

The ability of a primary protein structure to fold spontaneously to its native secondary or tertiary conformation, without any information other than the amino acid sequence and the noncovalent forces that act on the sequence, has been demonstrated. RNase will spontaneously refold to its native conformation after being denatured with loss of native structure but without the hydrolysis of peptide bonds. Such observations led to the hypothesis that a polypeptide sequence contains the properties sufficient to promote spontaneous protein folding to its unique active conformation under the correct solvent conditions and in the presence of prosthetic groups that may be a part of its structure. As described below chaperone proteins may facilitate the rate of protein folding. Quaternary structures also assemble spontaneously, after the tertiary structure of the individual polypeptide subunits are formed.

It may appear surprising that a protein folds into a single unique conformation given all the possible *a priori* rotational conformations available around single bonds in the primary structure. For example, the α-chain of hemoglobin contains 141 amino acids in which there are at least four single bonds per amino acid residue around which free rotation can occur. If each bond about which free rotation occurs has two or more stable rotamer conformations accessible to it, then there are a minimum of 4^{141} or 5×10^{86} possible conformations for the α-chain amino acid sequence.

The **conformation of a protein** is that conformation of the lowest Gibbs free energy accessible to the amino acid sequence within a physiological time frame. Thus folding is under thermodynamic and kinetic control. Although an exact knowledge of *de novo* folding of a polypeptide is at present an unattainable goal, certain processes appear reasonable. There is evidence that folding is initiated by short-range interactions forming secondary structures in small regions of the polypeptide. Short-range interactions are noncovalent interactions that occur between a side chain and its nearest neighbours. Particular side chains have a propensity to promote the formation of α-helices, *β*-structure, and sharp turns or bends (β-turns) in the polypeptide. The interaction of a side chain with its nearest neighbours in the polypeptide determines the secondary structure into which that section of the polypeptide strand folds. Sections of polypeptide, called **initiation sites,** thus spontaneously fold into small regions of secondary structure. The partially folded structures then condense with each other to form a **molten-globular** state. This is a condensed intermediate on the folding pathway that contains much of the secondary structure elements of the native structure, but a large number of incorrect tertiary structure interactions. Segments of secondary structure in the molten-globular state are highly mobile relative to one another, and the molten-globular structure is in rapid equilibrium with the fully unfolded denatured state. The correct medium- and long-range interactions between different initiation sites are found by rearrangements within the molten-globule and the low free energy, native tertiary structure for the polypeptide chain is formed. With formation of the native tertiary structure, the correct disulfide bonds (cystine) are formed. The rate-determining step for folding and unfolding of the native conformation lies between the molten-globular state and the native structure.

Cells contain proteins that facilitate the folding process. These include *cis-trans*-prolyl isomerases, protein disulfide isomerases, and chaperone proteins, ***cis-trans*-Prolyl isomerases** increase the rate of folding by catalyzing interconversion of *cis*- and *trans*-peptide bonds of proline residues within the polypeptide chain. This allows the correct proline peptide bond conformation to form for each proline as required by the folded native structure. **Protein disulfide isomerases** catalyze the breakage and formation of disulfide cystine linkages so incorrect linkages are not stabilized and the correct arrangement of cystine linkages for the folded conformation is rapidly achieved.

Chaperone proteins were discovered as **heat shock proteins (hsps),** a family of proteins whose synthesis is increased at elevated temperatures. The chaperones do not change the final outcome of the folding process but act to prevent protein aggregation prior to the completion of folding and to prevent formation of metastable dead-end or nonproductive intermediates during folding. They increase the rate of the folding process by limiting the number of unproductive folding pathways available to a polypeptide. Chaperones of the hsp 70-kDa family bind to polypeptide chains as they are synthesized on the ribosomes, shielding the hydrophobic surfaces that would normally be exposed to solvent. This protects the protein from aggregation until the full chain is synthesized and folding can occur. Some proteins, however, cannot complete their folding process while in the presence of hsp 70 chaperones and are delivered to the hsp60 family of chaperone proteins, also called **chaperonins.** The chaperonins form long cylindrical multisubunit quaternary structures that bind unfolded proteins in their molten-globular state within their central hydrophobic cavity. Chaperonins have an ATPase activity, hydrolyzing ATP as they facilitate folding. Chaperone proteins are also required for refolding of proteins after they cross cellular membranes. A system of chaperones exists to facilitate protein transport into mitochondria and into and through the endoplasmic reticulum. Proteins cross the lipid bilayer of the mitochondrial and endoplasmic reticulum membrances in an unfolded conformation, and local chaperones are required to facilitate their folding.

2.7 Hydrogen Bonds and Electrostatic Interactions

Hydrogen bonds are formed when a hydrogen atom covalently bonded to an electronegative atoms is shared with a second electronegative atom. The atom to which the hydrogen atom is covalently bonded is designated the hydrogen-**donor atom.** The atom with which the hydrogen atom is shared is the hydrogen-**acceptor atom.** Typical hydrogen bonds found in proteins are α-Helical and β-structure conformations are extensively hydrogen bonded.

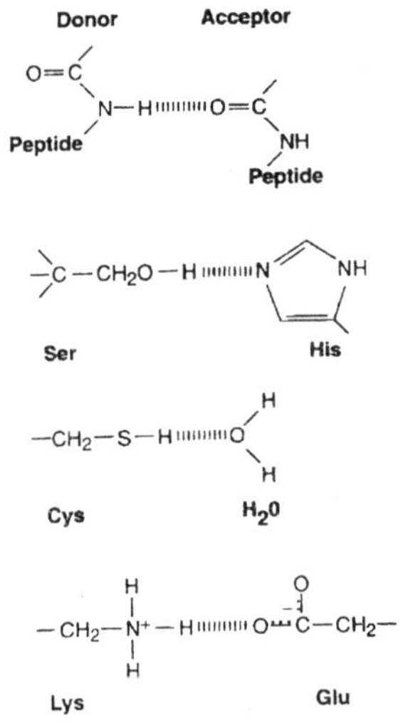

Donor Acceptor

O=C N–H ······ O=C NH

Peptide Peptide

–C–CH_2O–H ······ N NH

Ser His

–CH_2–S–H ······ O(H)(H)

Cys H_2O

–CH_2–N^+H_2–H ······ O^-–C(=O)–CH_2–

Lys Glu

Some common hydrogen bonds found in proteins.

The strength of a hydrogen bond is dependent on the distance between the donor and acceptor atoms. High bonding energies occur when the distance is between 2.7 and 3.1 A. Of lesser importance, but not negligible, to bonding strength is the dependence of hydrogen-bond strength on geometry. Bonds of higher energy are geometrically collinear, with donor, hydrogen, and acceptor atoms lying in a straight line. The dielectric constant of the medium around the hydrogen-bond may also be reflected in the bonding strength. Typical hydrogen-bond strengths in proteins are 1-7 kcal mol^{-1}. Although hydrogen bonds contribute to their modynamic stability of a protein's conformation, their formation may not be as major a driving force for folding as we might at first believe. This is because peptide bonds and other hydrogen-bonding groups in proteins form hydrogen bonds to the water solvent in the denatured state, and these bonds must be broken before the protein folds. The energy required to break the hydrogen bonds to water must be subtracted from the energy gained from formation of new hydrogen bonds between atoms in the folded protein in calculating the net contribution of hydrogen-bonding forces to the folding.

Electrostatic interactions (also referred to as **ionic** or **salt linkages**) between charged groups are important in the stabilization of protein structure and in binding of charged ligands and substrates to proteins. Electrostatic forces are repulsive or attractive depending on whether the interacting charges are of the same or opposite sign. The strength of an electrostatic force (E_{el}) is directly dependent on the charge (Z) of each ion and is inversely dependent on the dielectric constant (D) of the solvent and the distance between the charges (r_{ab}).

Water has a high dielectric constant ($D = 80$), and interactions in water are relatively weak in comparison to the strength of charge interactions in the interior of a protein where the dielectric constant is low. However, most charged groups of proteins remain on the surface of the protein where they do not interact with other charged groups from the protein because of the high dielectric constant of the water solvent, but are stabilized by hydrogen bonding and polar interactions to the water. These water interactions generate the dominant forces that lead to placement of most charged groups of a protein on the outside of the folded structures.

The distance of maximum favorable interaction between two atoms is the van der Waals contact distance, which is the sum of the van der Waals radii for the two atoms. The van der Waals repulsive forces between atoms attached to a peptide bond are weakest at the specific ϕ and ψ angles compatible with the α-helix and β-strand structures. Thus van der Waals forces are critical for secondary structure formation in proteins. In folding into a tertiary structure, the number of weak van der Waals interactions that occur are in the thousands. Thus the total contribution of van der Waals-London dispersion forces to the stability of a folded structure is substantial, even though a single interaction between any two atoms is less than 1 kcal mol^{-1}.

A special type of interaction (π-electron–π-electron) occurs when two aromatic rings approach each other with electrons of their aromatic rings favorably interacting. This interaction can result in attractive forces of up to 6 kcal mol^{-1}. A number of π-π aromatic interactions occur in a typical folded protein, contributing to the stability of the folded structure.

2.8 Aspects of Protein Structure

While high-resolution X-ray diffraction experiments yield atomic coordinates for each atom in a protein structure, experimental evidence from NMR, fluorescence spectroscopy, and the temperature dependence of the X-ray diffraction reveals that the atoms in a folded protein molecule have a fluid-like dynamic motion and do not exist in a single static position. Rather than an exact location, the atomic coordinates obtained by X-ray diffraction represent the time-averaged position for each atom. The time frame for position averaging is the length of time for data collection, which may be several days. Thus the active conformation may differ from the average conformation. An X-ray structure also shows small "defects" in packing of the folded structure, indicating the existence of "holes" in the structure that will allow the protein space for flexibility. The concept that each atom in a protein is in constant motion such as molecules within a fluid, although constrained by its covalent bonds and the secondary and tertiary structure, is an important aspect of protein structure.

Theoretical **molecular dynamics** calculations describe the changes in coordinates of atoms in a folded protein structure, with corresponding changes in position of regions of the structure due to summation of the movements of atoms in that region. The dynamic motion computation is based on the solving of Newton's equations of motion simultaneously for all the atoms of the protein, and the solvent that interacts with the protein. The energy functions used in the equation include representations of covalent and noncovalent bonding energies due to electrostatic forces, hydrogen bonding, and van der Waals forces.

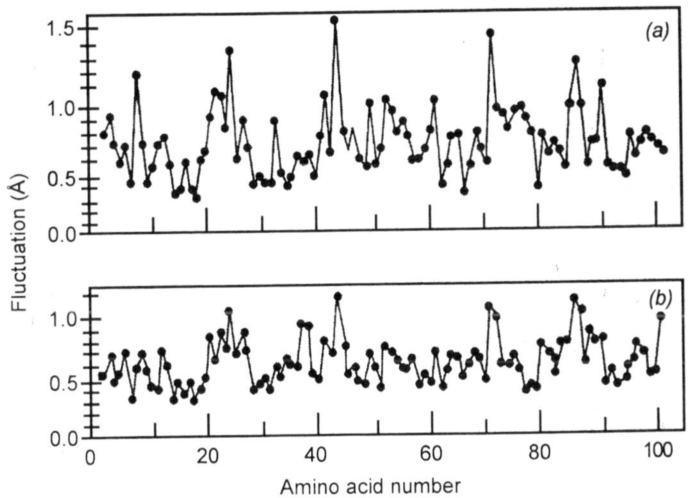

Fluctuation of structure of cytochrome *c*.

Individual atoms are randomly assigned a velocity from a theoretical distribution and Newton's equations are used to "relax" the system at a given "temperature". The calculation is a computationally intensive activity, even when limited to less than several hundred picoseconds (1 ps = 10^{-12} s) of protein dynamic time, and frequently requires supercomputers. These calculations indicate that the average atom within a typical protein is oscillating over a distance of 0.7 Å on the picosecond scale. Some atoms or groups of atoms move smaller or larger distances than this calculated average.

Net movement of any segment of a polypeptide over time represents the sum of forces due to rapid atomic oscillations and the local jiggling and elastic movements of covalently attached groups of atoms. These movements within the closely packed interior of a protein molecule are frequently large enough to allow the planar aromatic rings of buried tyrosines to flip. Furthermore, the small amplitude fluctuations provide the "lubricant" for large motions in proteins such as domain motions and quaternary structure changes, like those observed in hemoglobin on O_2 binding. The dynamic behaviour of proteins is implicated in conformational changes induced by substrate, inhibitor, or drug binding to enzymes and receptors, generation of allosteric effects in hemoglobins, electron transfer in cytochromes, and in the formation of supramolecular assemblies such as viruses. The movements may also have a functional role in the protein's mechanism of action.

2.9 Determination of the Amino Acid Sequence

The ability to clone genes for proteins has led to the determination of the amino acid sequence of a protein as derived from the DNA or RNA sequences. This is a much faster method for obtaining an amino acid sequence. Sequencing of a protein, however, is required for the determination of modifications to the protein structure that occur after its biosynthesis, to identify a part of the protein sequence in order that its gene can be cloned, and to identify a protein as the product of a particular gene. Determination of the primary structure of a protein requires a purified protein. Many proteins contain several polypeptide chains and it is necessary to determine the number of chains in the protein. Individual chains are purified by the same techniques used in purification of the whole protein. If disulfide bonds covalently join the chains, these bonds have to be broken.

Polypeptide chains are most commonly sequenced by the **Edman reaction** in which the polypeptide chain is reacted with phenylisothiocyanate, which forms a covalent bond to the NH_2-terminal amino acid. In this derivative, acidic conditions catalyze an intramolecular cyclization that results in cleavage of the NH_2-terminal amino acid from the polypeptide chain as a phenylthiohydantoin derivative. This NH_2-terminal amino acid derivative may be separated chromatographically and identified against standards. The polypeptide chain minus the NH_2-terminal amino acid is then isolated.

Edman reaction is repeated to identify the next NH_2-terminal amino acid. This series of reactions can theoretically be repeated until the sequence of the entire polypeptide chain is determined but under favourable conditions can only be carried out for 30 or

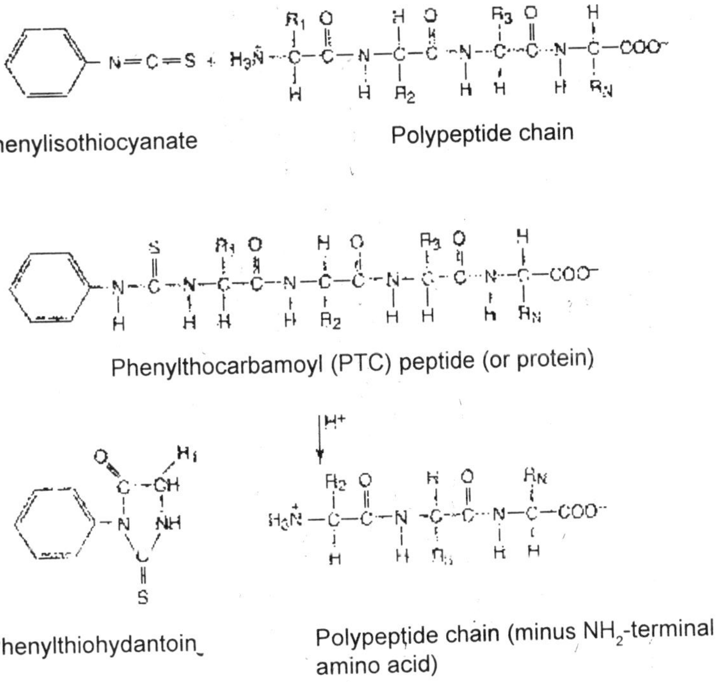

Phenylisothiocyanate

Polypeptide chain

Phenylthocarbamoyl (PTC) peptide (or protein)

H^+

Phenylthiohydantoin

Polypeptide chain (minus NH_2-terminal amino acid)

Edman reaction.

40 amino acids into the polypeptide chain. At this point in the analysis, impurities generated from incomplete reactions in the reaction series make further **Edman cycles** unfeasible. Since most polypeptide chains in proteins contain many more than 30 or 40 amino acids, they have to be hydrolyzed into smaller fragments and sequenced in sections.

Enzymatic and chemical methods are used to break polypeptide chains into smaller polypeptide fragments for sequencing. The enzyme **trypsin** preferentially catalyzes hydrolysis of the peptide bond on the COOH-terminal side of the basic amio acid residues of lysine and arginine within polypeptide chains. **Chymotrypsin** hydrolyzes peptide bonds on the COOH-terminal side of residues with large apolar side chains. Other proteolytic enzymes cleave polypeptide chains on the COOH-terminal side of glutamic and aspartic acid. **Cyanogen bromide** specifically cleaves peptide bonds on the COOH-terminal side of methionine residues within polypeptide chains.

To establish the amino acid sequence of a large polypeptide, it is subjected to partial hydrolysis by one of the specific cleaving reagents, the polypeptide segments are separated and the amino acid sequence of each of the small segment is, determined by the Edman reaction.

To place the sequenced peptides correctly into the complete sequence of the original polypeptide, a sample of the original polypeptide is subjected to a second partial hydrolysis, by a different hydrolytic reagent from that used initially. This generates overlapping sequences to the first set of sequences, leading to the complete sequence.

2.10 Evaluation of Protein Structure

Ultraviolet Light Spectroscopy

The side chains of tyrosine, phenylalanine, and tryptophan, as well as peptide bonds in proteins, absorb **ultraviolet (UV) light.** The efficiency of light energy absorption for each **chromophore** is related to its **molar extinction coeffient** (ε). The absorbance between 260 and 300 nm is primarily due to phenylalanine, tyrosine, and tryptophan side chain groups. When the tyrosine side chain is ionized at high pH (the tyrosine R group has a $pK_a \approx 10$), the absorbance for tyrosine is shifted to a higher wavelength (red shifted) and its molar absorptivity is increased. Peptide bonds absorb in the far-UV (180–230 nm). A peptide bond in α-helix conformation interacts with the electrons of other peptide bonds above and below it in the spiral conformation to create an **exciton system** in which electrons are delocalized. The result is a shift of the absorption maximum from that of an isolated peptide bond to either a lower or higher wavelength. Thus UV spectroscopy can be used to study changes in a protein's secondary and tertiary structure. As a protein is denatured (helix unfolded), differences are observed in the absorption characteristics of the peptide bonds due to the disruption of the exciton system. In addition, the absorption maximum for an aromatic chromophore appears at a lower wavelength in an aqueous environment than in a nonpolar environment.

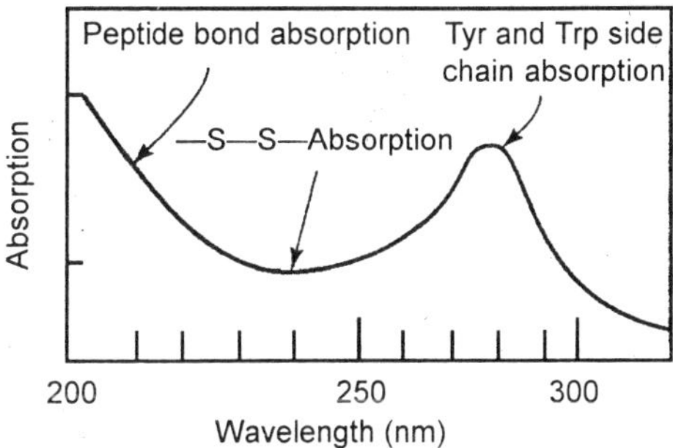

Ultraviolet absorption spectrum of the globular protein α-chymotrypsin.

The **molar absorbency** of a chromophoric substrate often changes on binding to a protein. This change in the binding molecule's extinction coefficient can be used to measure its binding constant. Changes in chromophore extinction coefficients during the enzyme catalysis of a chemical reaction are used to obtain the kinetic parameters for the reaction.

By using Fluorescence spectroscopy protein structure, can be evaluated successfully. The things what we have to remember is that, the fluorescent emission is always at a longer wavelength of light than the absorption wavelength of the fluorophore.

Nuclear Magnetic Resonance

With **two-dimensional** (2-D) **NMR** and powerful NMR spectrometers it is possible to obtain the solution conformation of small proteins of approximately 150 amino acids or less. Multidimensional NMR and triple resonance can extend the NMR to solve protein structures with up to 250 amino acids.

Conventional NMR techniques involve use of radiofrequency (rf) radiation to study the environment of atomic nuclei that are magnetic. The requirement for magnetic nuclei is absolute and is based on an unpaired spin state in the nucleus. Thus the naturally abundant carbon (^{12}C), nitrogen (^{14}N), and oxygen (^{16}O) do not absorb, while ^{13}C, ^{15}N, and ^{17}O do absorb. The absorption bands in an NMR spectrum are characterized by (1) a position or chemical shift value, reported as the frequency difference between that

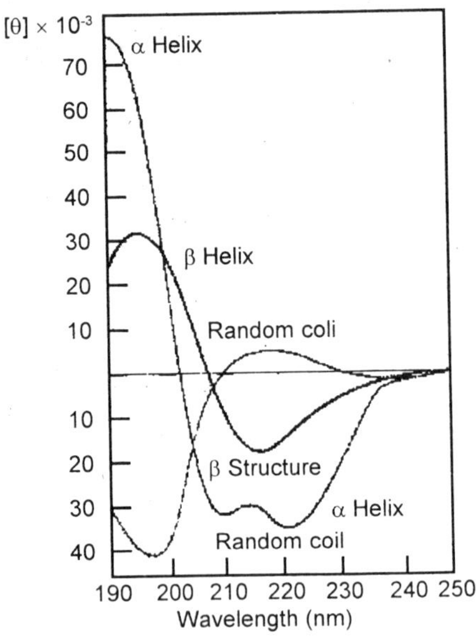

Circular dichroism spectra for polypeptide chains in α-helix, β-structure, and random coil conformations.

observed for a specific absorption band and that for a standard reference material; (2) the intensity of the peak or integrated area, which is proportional to the total number of absorbing nuclei; (3) the half-height peaks width, which reflects the degree of motion in solution of the absorbing species; and (4) the coupling constant, which measures the extent of direct interaction or influence of neighbouring nuclei on the absorbing nuclei. These four measurements enable the determination of the identity and number of nearest-neighbour groups that can affect the response of absorbing species through bonded interactions. They give no information on through-space (nonbonded) interaction due to the three-dimensional structure of the protein. To determine through-space interactions and protein tertiary structure requires the use of **Nuclear Overhauser Effects** (NOEs) and the application of the two-dimensional technique.

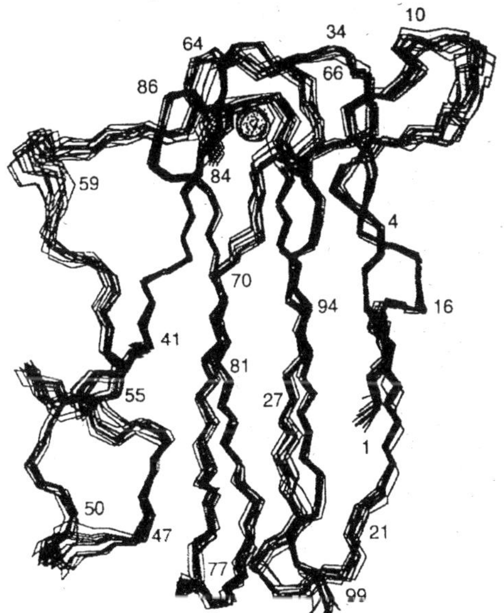

NMR structure of the protein plastocyanin from the French bean.

The major difference in two-dimensional versus one-dimensional (1-D) NMR is the addition of a second time delay rf pulse. The technique first requires the identification in the spectrum of a proton absorbance from a particular position in the protein structure. A maximum distance of approximately 5 Å is the limit for which these through-space interactions can be observed. Upon he generation of distance information for interresidue pairs through the protein structure, three-dimensional protein conformations consistent with the spectra are generated. In this calculation, a distance matrix is constructed containing ranges of distances (minimum and maximum) for as many interresidue interactions as may be measured. Possible structures are generated from the data consistent with the constraints imposed by the NMR spectra. Computational refinements of the initially calculated structures can be made to optimize covalent bond distances and bond angles. The method generates a family of structures, the variability showing either the imprecision of the technique or the dynamic "disorder" of the folded structure. Such computations based on NMR experiments have yielded structures for proteins that do not significantly differ from the time-averaged structure observed with X-ray diffraction methods.

3

Specific Protein Families

3.1 Introduction

The specific relationships between structure and function in four protein families can be studied as follows.

1. Immunoglobulin family
2. The serine proteases
3. DNA-binding proteins
4. The hemoglobin family.

Now let us examine the specific relationships between structure and function in four **protein families:** immunoglobulins, serine proteases, DNA-binding proteins, and hemoglobins. We pursue this study through the examination of the variability in amino acid sequence, structural organization, and biological function. The significance of the structure-function relationship can best be appreciated through observation of the range of such variations within specific protein families.

The **immunoglobulin family** provides examples of multidomain architecture that supports recognition and binding to foreign molecules and leads to their sequestration. Diversity among family members is the source of specific molecular recognition and individual binding capabilities.

The **serine proteases** provide examples of a family of enzymes that appear to have diverged to perform unique physiological functions, frequently highly organized within enzyme cascade processes. Their inherent similarities in catalytic mechanism and three-dimensional structure are a common link.

DNA-binding proteins are multifamilies of proteins that bind to regulatory sites in DNA and regulate gene expression, an amazing feat as the mammalian genome contains approximately 100,000 unique genes. These proteins contain unusual super-secondary structure motifs that allow them to selectively bind regulatory sites of specific genes.

The **hemoglobin** family offers examples of a highly fine-tuned system that can accommodate small substitutions or mutations, many of which have been studied as to their clinical implications. This family reveals the potential diversity of amino acid sequence substitutions that can be tolerated and allow the protein to function in an acceptable physiological manner.

While studying about Immunoglobulin it is very much needed to understand about antibody molecules.

3.2 Immunoglobulin Family

Antibody molecules are produced in response to invasion by foreign compounds that can be proteins, carbohydrates, and nucleic acid polymers. An antibody molecule noncovalently associates with the foreign substance, initiating a process by which the foreign substance can be eliminated from the organism.

Molecules that induce antibody production are **antigens** and may contain multiple antigenic determinants, small regions of the antigen molecule that elicit the production of a specific antibody to which the antigen binds. In proteins, an antigenic determinant may comprise only six or seven amino acids.

A **hapten** is a small molecule that cannot alone elicit production of specific antibodies but when covalently attached to a larger molecule it acts as an antigenic determinant and induces antibody synthesis. Whereas hapten molecules need attachment to a larger molecule to elicit antibody synthesis, when detached from their carrier, they will retain the ability to bind strongly to antibody.

It is estimated that each human can potentially produce about 1×10^8 different antibody structures. All antibodies, however, have a similar structure. The determination of the structure has been accomplished from studies of immunoglobulin primary structures and X-ray diffraction that show the three-dimensional structure of the antibody molecule alone and in complex with antigen.

Structural studies of proteins require pure homogeneous preparations. Such samples of antibodies are extremely difficult to isolate from blood because of the wide diversity of antibody molecules present. Homogeneous antibodies can be obtained, however, by the monoclonal hybridoma technique in which mouse myeloma cells are fused with mouse antibody-producing B lymphocytes to construct immortalized hybridoma cells that express a single antibody.

Now let us learn something about the functions of different Antibody classes.

The IgA class of immunoglobulins is found primarily in the mucosal secretions (bronchial, nasal, and intestinal mucous secretions, tears, milk, and colostrum). These immunoglobulins are the initial defense against invading viral and bacterial pathogens prior to their entry into plasma or other internal space.

The IgM class is found primarily in plasma. They are the first antibodies elicited in significant quantity on the introduction of a foreign antigen into a host's plasma. IgM antibodies promote phagocytosis of microorganisms by macrophage and polymorpho-nuclear leukocytes and are also potent activators of complement. IgM antibodies occur in many external secretions but at levels lower than those of IgA.

The IgG class occurs in high concentration in plasma. Their response to foreign antigens takes a longer period of time than that of IgM. At maximum concentration they are present in significantly higher concentration than IgM. Like IgM, IgG antibodies promote phagocytosis in plasma and activate complement.

The normal biological functions of the IgD and IgE classes of immunoglobulins are not known; however, the IgE antibodies play an important role in allergic responses such as anaphylactic shock, hay fever, and asthma.

Immunoglobulin deficiency usually causes increased susceptibility to infection. X-linked agammaglobulinemia and common variable immunodeficiency are two examples. The commonest disorder is selective IgA deficiency, which results in recurrent infections of sinuses and the respiratory tract.

Immunization: An immunizing vaccine can consist of killed bacterial cells, inactivated viruses, killed parasites, a nonvirulent form of live bacterium related to a virulent bacterium, a denatured bacterial toxin of recombinant protein. The introduction of a vaccine into a human can lead to protection against virulent forms of microorganisms or toxic agents that contain the same antigen. Antigens in nonvirulent material not only cause the differentiation of lymphoid cells so that they produce antibody toward the foreign antigen but also cause differentiation of some lymphoid cells into memory cells. Memory cells do not secrete antibody but place antibodies to the antigen onto their outer surface, where they act as future sensors for the antigen. These memory cells are like a longstanding radar for the potentially virulent antigen. On reintroduction of the antigen at a later time, the binding of the antigen to the cell surface antibody in the memory cells stimulates the memory cell to divide into antibody-producing cells as well as new memory cells. This reduces the time for antibody production that is required on introduction of an antigen and increases the concentration of antigen-specific antibody produced. It is the basis for the protection provided by immunization.

Recently introduced vaccines for adults include pneumococcal vaccine (to prevent pneumonia due to *Diplococcus pneumoniae*), hepatitis B vaccine, and influenza vaccine. The latter changes each year to account for antigenic variation in the influenza virus.

Now let us study in brief about immunoglobulin fold.

With in each of the polypeptide chains of an antibody molecule is a repeating pattern of amino acid sequences.

Each of the 110 amino acid segments form separate structural domains of similar tertiary structure as shown by X-ray diffraction studies. Each 110 segment of the H and L chains folds into a supersecondary structure with a unique but similar arrangement of antiparallel β-strands, which generates a motif known as an **immunoglobulin fold**. This motif consists of 7 to 9 polypeptide strands that form two antiparallel β-sheets that are aligned face-to-face. Globular domains result from the strong interaction between two immunoglobulin folds on two separate chains. The associations are between domains V_L–V_H and C_L–C_H1 in the H and L chains. In the C-terminal half of the H chains, the two chains associate to generate domains C_H2–C_H2 and C_H3–C_H3. A "hinge" polypeptide sequence interconnects the two C_L–C_H1 domains with the C_H2–C_H2 domain in the antibody structure.

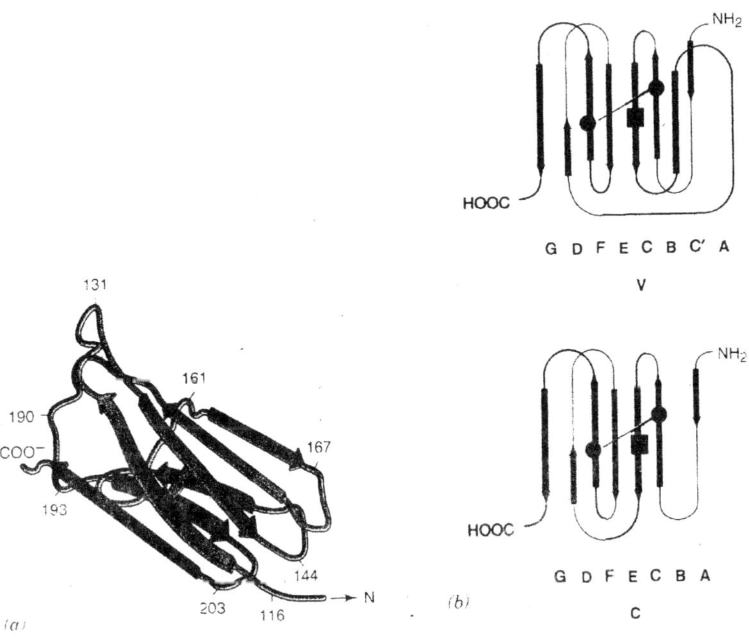

Immunoglobulin fold.

(a) Schematic diagram of folding of a C_L domain, showing β-pleated sheet structure. Arrows show strands of β-sheet and bar (blue) shows position of cystine bond. Light arrows are for β-strands in plane above and dark arrows are β-strands in plane below. (b) Diagrammatic outline of arrangement of β-strands in immunoglobulin fold motif. Examples are for IgG variable and constant regions. Thick arrows indicate β-strands and thin lines loops that interconnect the β-strands. Circles indicate cysteines that form intradomain disulfide bond. Squares show positions of tryptophan residues that are an invariant component of the core of the immunoglobulin fold. Boldface black letters indicate strands that form one plane of the sheet, while other letters form the parallel plane behind the first plane.

3.3 The Serine Proteases

Serine proteases are a family of enzymes that utilize a single uniquely activated serine residue in their substrate-binding site to catalytically hydrolyze peptide bonds. This serine can be characterized by the irreversible reaction of its side chain hydroxyl group with diisopropylfluorophosphate of all the serines in the protein. DFP reacts only with the catalytically active serine to form a phosphate ester.

Now let us understand about the involvement of serine proteases in tumor cell metastasis.

The serine protease urokinase is believed to be required for the metastasis of cancer cells. Metastasis is the process by which a cancer cell leaves a primary tumor and migrates through the blood or lymph system to a new tissue or organ, where a secondary tumor grows. Increased synthesis of urokinase has been correlated with an increased ability to metastasize in many cancers. Urokinase activates plasminogen to form plasmin. Plasminogen is ubiquitously located in extracellular space and its activation to plasmin can cause the catalytic degradation of the proteins in the extracellular matrix through which the metastasizing tumor cells migrate. Plasmin can also activate procollagenase to collagenase, promoting the degradation of collagen in the basement membrane surrounding the capillaries and lymph system. This promotion of proteolytic degradative activity by the urokinase secreted by tumor cells allows the tumor cells to invade the target tissue and form secondary tumor sites.

Some Serine Proteases and their Biochemical and physiological roles can be understood as follows.

Protease	*Action*	*Possible Disease Due to Deficiency or Malfunction*
Plasma kallikrein Factor XIIa Factor XIa	Coagulation (see Clin. Corr. 3.4)	Cerebral infarction (stroke) coronary infarction, thrombosis, bleeding disorders
Factor IXa		
Factor VIIa		
Factor Xa		
Factor IIa (thrombin)		
Activated protein C		
Factor Clr Factor Cls	Complement (see Clin. Corr. 3.1)	Inflammation, rheumatoid arthritis, autoimmune disease
Factor D		
Factor B		
C3 convertase		
Trypsin	Digestion	Pancreatitis
Chymotrypsin		
Elastase (pancreatic)		
Enteropeptidase		
Urokinase plasminogen activator Tissue plasminogen activator	Fibrinolysis, cell migration, embryogenesis, menstruation	Clotting disorders, tumor metastasis (see Clin. Corr. 3.5)
Plasmin		
Tissue kallikreins	Hormone activation	
Acrosin	Fertilization	Infertility
α–Subunit of nerve growth factor	Growth factor activation	
γ–Subunit of nerve growth factor		
Granulocyte elastase Cathepsin G Mast cell chymases	Extracellular protein and peptide degradation, mast cell function	Inflammation, allergic response
Mast cell tryptases		

Serine proteases are synthesized in an inactive **zymogen** form, which requires limited proteolysis to produce the active enzyme. Those for coagulation are synthesized in liver cells and are secreted into the blood for subsequent activation by other serine proteases following vascular injury. Zymogen forms are usually designated by the suffix *-ogen* after the enzyme name; the zymogen form of trypsin is termed trypsin*ogen* and for chymotrypsin is termed chymotrypsin*ogen*. In some cases the zymogen form is referred to proenzyme; the zymogen form of thrombin is prothrombin.

Several plasma serine proteases secrete zymogen forms that contain **multiple nonsimilar domains.** Protein C, involved in a fibrinolysis pathway in blood, has four distinct domains. The NH_2-terminal domain contains the derived amino acid, **γ-carboxyglutamic acid,** which is enzymatically formed by carboxylation of glutamic acid residues in a vitamin K-dependent reaction. The γ-carboxyglutamic acids chelate calcium ions and form part of a binding site to membrances. The COOH-terminal segment contains the catalytic domains. Activation of these zymogens requires specific proteolysis outside the catalytic domains and is controlled by the binding through the nine γ-carboxyglutamic acid residues at the NH_2-terminal end to a membrane.

Evolutionary selection of this enzyme family for participation in physiological processes requires a parallel evolution of control factors. Specific proteins inhibit the activities of serine proteases after their physiological role has ended. Thus coagulation is limited to the site of vascular injury and complementation activation leads to lysis only of cells exhibiting foreign antigens. Inability to control these protease systems, which may be caused by a deficiency of a specific inhibitor, can lead to undesirable consequences, such as thrombi formation in myocardial infarction and stroke or uncontrolled reactions of complement in autoimmune disease. Natural inhibitors of serine proteases, termed **serpins** for serine protease inhibitors, have evolved. This family of inhibitors occurs in animals that have the proteases, but surprisingly these inhibitors are also found in plants that lack proteases.

Some Human Proteins that Inhibit Serine Proteases

Inhibitor	*Action*
α_1-Proteinase inhibitor	Inhibits tissue proteases including neutrophil elastase; deficiency leads to pulmonary emphysema
α_1-Antichymotrypsin	Inhibits proteases of chymotrypsin-like specificity from neutrophils, basophils, and mast cells including cathepsin G and chymase
Inter-α-trypsin inhibitor	Inhibits broad range of serine protease activities in plasma
α_2-Antiplasmin	Inhibits plasmin
Antithrombin III	Inhibits thrombin and other coagulation proteases
C_1 Inhibitor	Inhibits complement reaction
α_2-Macroglobulin	General protease inhibitor
Protease nexin I	Inhibits thrombin, urokinase, and plasmin
Protease nexin II	Inhibits growth factor-associated serine proteases, identical to H_2N-terminal domain of amyloid protein secreted in Alzheimer's disease
Plasminogen activator inhibitor I	Inhibits plasminogen activators
Plasminogen activator inhibitor II	Inhibits urokinase plasminogen activator

The complex relationships between structure and physiological function in the serine proteases require analysis of a number of observations. (1) Only one serine residue is catalytically active and participates in peptide bond hydrolysis. Bovine trypsin contains 34 serine residues with only one catalytically active or able to react with the inhibitor DFP. (2) X-ray diffraction and amino acid sequence homology studies demonstrates that there are two residues, a histidine and an aspartate, that are always associated with the activated serine in the catalytic site.

Much of our early knowledge of the serine protease family came from trypsin and chymotrypsin purified from bovine materials obtained from a slaughter-house. This has yielded a useful but nonintuitive nomenclature, which uses a sequence alignment against the amino acid sequence of chymotrypsin, to name and number residues of other serine proteases. As mentioned previously, the catalytically essential residues are Ser 195, His 57, and Asp 102. Insertions and deletions of the amino acids in another serine protease are compared to the numbering of residues in chymotrypsin. Alignment is made by algorithms that maximize sequence homology, with exact alignment of the essential serine, histidine, and aspartate residues. These three residues are invariant in all serine proteases and the sequences surrounding them are invariant among the serine proteases of the chymotrypsin family.

Members of the chymotrypsin family also occur in prokaryotes. Thus bacterial serine proteases from *Streptomyces griseus* and *Myxobacteria* 450 have a structural and functional homology with chymotrypsin. A separate class of serine protease enzymes has been isolated, however, from bacteria that has no structural homology to the mammalian chymotrypsin family. The serine protease subtilisin, isolated from *Bacillus subtilis,* hydrolyzes peptide bonds and contains an activated serine with a histidine and aspartate in its active site but the active site arises from structural regions of the protein that bear no sequence or structural homology with the chymotrypsin serine proteases. This serine protease is an example of **convergent evolution** of an enzyme catalytic mechanism. Apparently a gene completely different from those that code for chymotrypsin-like serine proteases evolved the same catalytic mechanisms utilizing an active-site serine. The primary and tertiary structure, however, is different from that of the trypsin- and chymotrypsin-like structure.

3.4 DNA-Binding Proteins

Regulatory sites exist in DNA that bind proteins that control gene expression. These sites contain a nucleotide sequence that binds regulatory proteins known as transcription factors. The specific DNA sequence, or **transcription factor binding element,** is usually less than 10 nucleotides long. Noncovalent interactions between the protein and DNA allow the protein to recognize the nucleotide sequence and bind to a specific regulatory site. This is a highly selective feat as the human genome has up to 100,000 genes, each with its own regulatory sequences. While there are huge gaps in our knowledge of how proteins regulate gene expression, some common structural motifs of DNA-binding proteins are apparent.

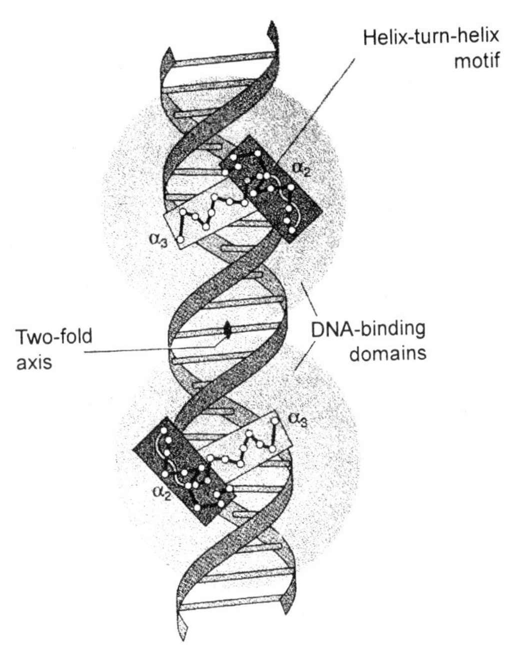

Association of a DNA-binding protein (dimer) with two helix-turn-helix motifs into adjacent major grooves of B-DNA.

Along the helical spiral of a DNA molecule in its most common form (B form) are two grooves, the major and minor grooves to which the proteins must associate. A structural motif found in many DNA-binding proteins is the **helix-turn-helix (HTH).** An HTH places one of its α-helices, designated the **recognition helix,** across the major groove where side chain residues of the helix form specific noncovalent interactions with the base sequence of the target DNA. The interaction appears to induce distortions in conformation of the B-DNA binding site that better accommodate the interactions with protein. Nonspecific interactions are made between the protein and sugar-phosphate backbone of DNA. HTH proteins bind as dimers; thus there are two helix-turn-helix motifs per active regulatory protein. X-ray structures show the two helix-turn-helix motifs protruding from the structure of each monomer domain binding at two adjacent turns of the major groove in the DNA, making a strong protein–DNA interaction.

The **zinc-finger** motif is another structure found in some DNA-binding proteins. Zinc-finger proteins contain repeating motifs of a Zn^{2+} atom bonded to two cysteine and two histidine side chains. In some cases the histidines may be substituted by cysteines. The primary structure for the motif contains two close cysteines separated by about 12 amino acids from a second pair of Zn^{2+} liganding amino acids (histidine or cysteine). The three-dimensional structure of one zinc finger has been deduced by ^{1}H-NMR. The motif contains an α-helix segment that can bind within the major groove at its target site in DNA and makes specific interactions with the nucleotide base sequence.

A third structural motif found in some of the DNA-binding proteins is the **leucine zipper.** Leucine zippers are formed from a region of α-helix that contain at least four leucines, each leucine separated by six amino acids from one another (i.e., Leu-X_6-Leu-H_6-Leu-X_6Leu, where X is any common amino acid). With residues per turn of the α-helix, the leucines align on one edge of the helix, with a leucine at every second turn of the helix. The leucine-rich helix forms a hydrophobic interaction with a second leucine helix on another polypeptide chain subunit, to "zipper" the two subunits together to form a dimer. The leucine-zipper motif does not directly interact with the DNA, as do the zinc-finger or helix-turn-helix motifs. Mutations in the zipper motif show that if the dimer is not formed by association of the monomers through the zipper, the protein will not bind to DNA strongly. However, just adjacent to the α-helix of the zipper motif in the primary structures there is a sequence containing a high concentration of basic amino acids, arginine and lysine. This evolutionary conserved basic region interacts with the DNA. The positive charges of the arginine and lysine side chains are drawn to the negatively charged DNA phosphate groups.

The yeast transcription factor GCN4 is one eukaryotic DNA-binding protein that contains the leucine-zipper (bZIP) motif. It is a dimer of two continuous α-helical subunits joined by a leucine-zipper interface. The α-helices cross at this interface and then diverge with their two N-terminal ends separated to pass directly through different sides of the same major groove of the DNA target site. Amazingly, there are no bends or kinks in the linear helical structure of each subunit of the dimer. As discussed above, the DNA contact regions contain many positively charged amino acid residues that interact with negatively charged phosphate groups in the DNA.

Many regulatory proteins with the leucine-zipper motif have been shown to be oncogene products (Myc, Jun, and Fos). Fos forms a heterodimer with Jun through a leucine-zipper interaction, and the Fos/Jun dimers bind to gene regulatory sites. If these regulatory proteins are mutated or produced in an unregulated manner, the cell can be transformed to a cancer cell.

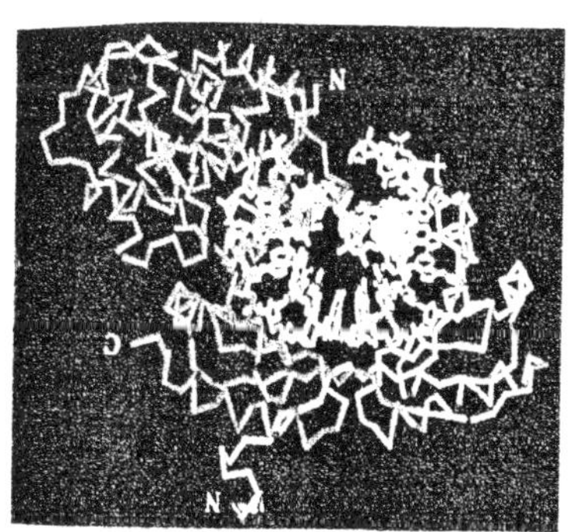

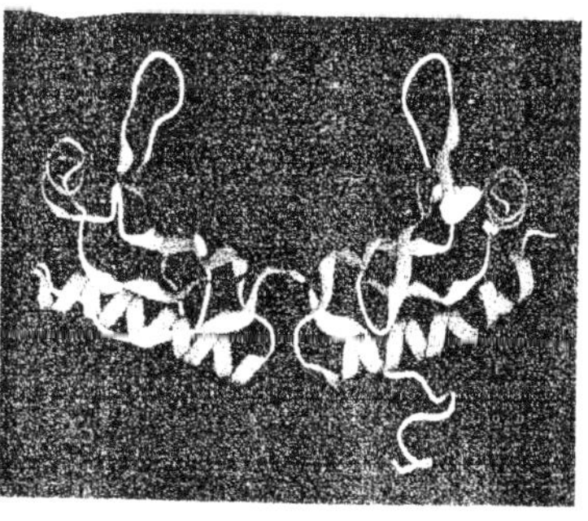

Structures of TBP-DNA binary and TBP-TFIIB-DNA ternary complexes.

The helix-loop-helix motif was the first motif to be identified for interaction with DNA. X-ray structural studies of protein–DNA complexes show a great variety of other mechanisms for protein–DNA association. The TATA box-binding protein (TBP) associates with the TATA sequence of gene promoters. Association of TBP with the TATA sequence forms the foundation for a large protein complex that initiates gene transcription by RNA polymerase. The X-ray structure of the C-terminal domain of the TBP bound to a TATA sequence shows that TBP contains two domains, each composed of a curved antiparallel β-sheet with a concave surface. The two-domain structure forms the shape of a "saddle" that sits over the DNA double helix. The concave surface of the "saddle" distorts the B-DNA structure and partially unwinds the DNA helix. This distortion, in turn, produces a wide open, though shallow, minor groove that interacts extensively with the under portion of the TBP saddle. One critical protein that forms a part of the initiation complex for RNA transcription is TFIIB. An X-ray structure shows TFIIB associates with one of the "stirrups" of the TBP "saddle" in the TATA sequence complex.

3.5 The Hemoglobin Family

Hemoglobins are globular proteins, present in high concentrations in red blood cells, that bind oxygen in the lungs and transport the oxygen in blood to tissues and cells around the capillary beds of the vascular system. Hemoglobins also transport carbon dioxide and protons from the tissues to the lungs. Hemoglobins carry and release nitric oxide (NO), a potent vasodilator and inhibitor of platelet aggregation. In this section the structural and molecular aspects of hemoglobin and myoglobin are described.

A hemoglobin molecule consists of four polypeptide chains, two each of two different amino acid sequences. The major form of human adult hemoglobin, **HbA$_1$**, consists of two α chains and two β chains ($\alpha_2\beta_2$). The α polypeptide has 141 and the β polypeptide has 146 amino acids. Other forms of hemoglobin predominate in the blood of the human fetus and early embryo. The fetal form **(HbF)** contains the same α chains found in HbA$_1$, but a second type of chain (γ chain) occurs in the tetramer molecule and differs in amino acid sequence from that of the β chain of adult HbA_1. Additional forms appear in the first months after conception (embryonic) in which the α chains are substituted by *zeta* (ζ) chains of different amino acid sequence and the ε chains serve as the β chains. A minor form of adult hemoglobin, HbA_2, comprises about 2% of normal adult hemoglobin and contains two α chains and two chains designated delta (δ).

Myoglobin (Mb) is an O_2-carrying protein that binds and releases O_2 with changes in the oxygen concentration in the sarcoplasm of skeletal muscle cells. In contrast to hemoglobin, which has four polypeptide chains and four O_2-binding sites, myoglobin contains only a single polypeptide chain and one O_2-binding site. Myoglobin is a model for what occurs when a single protomer molecule acts alone without the interactions exhibited among the four O_2-binding sites in the more complex tetramer molecule of hemoglobin.

The four polypeptides of globin subunits in hemoglobin and the one of myoglobin each contain a heme prosthetic group. A **prosthetic group** is a nonpolypeptide moiety that forms a functional part of a protein. Without its prosthetic group, a protein is designated an **apoprotein.** With its prosthetic group it is a **holoprotein**.

Heme contains protoporphyrin IX with an iron atom in its center. The iron atoms is in the ferrous (2+ charge) oxidation state in functional hemoglobin and myoglobin. The ferrous atoms in the heme can form five or six ligand bonds, depending on whether or not O_2 is bound to the molecule. Four bonds are to the pyrrole nitrogen atoms of the porphyrin. Since all pyrrole rings of porphyrin lie in a common plane, the four ligand bonds from the porphyrin to the iron atom will have a tendency to lie in the plane of the porphyrin ring. The fifth and the potentially sixth ligand bonds to the ferrous atom are directed along an axis perpendicular to the plane of the porphyrin ring. The fifth coordinate bond of the ferrous atom is to a nitrogen of a histidine imidazole. This is designated the **proximal histidine** in hemoglobin and myoglobin structures. O_2 forms a sixth coordiante bond to the ferrous atom when bound to hemoglobin. In this bonded position the O_2 is placed between the ferrous atom to which it is liganded and a second histidine imidazole, designated the **distal histidine.** In deoxyhemoglobin, the sixth coordination position of the ferrous atom is unoccupied.

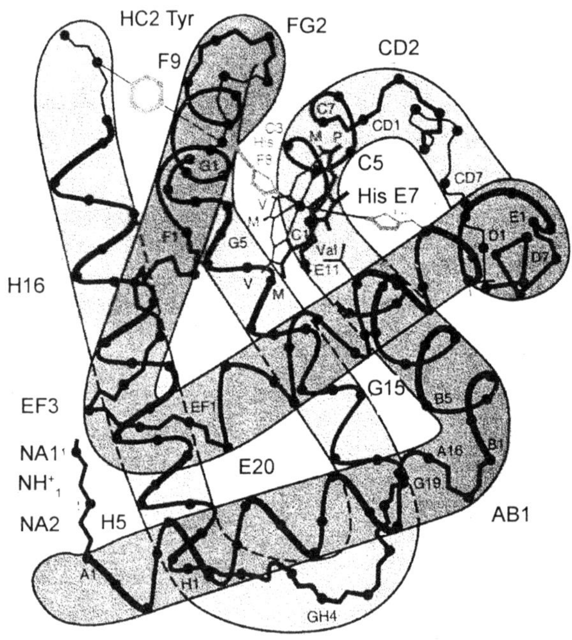

Secondary and tertiary structure characteristics of chains of hemoglobin.

X-ray diffraction data on deoxyhemoglobin show that the ferrous atoms actually sit out of the plane of their porphyrins by about 0.4-0.6 Å. This is thought to occur because of two factors. The electronic configuration of the five-coordinated ferrous atom in deoxyhemoglobin has a slightly larger radius than the distance from the center of the porphyrin to each of the pyrrole nitrogen atoms.

Accordingly, the iron can be placed in the center of the porphyrin only with some distortion of the porphyrin conformation. Probably a more important consideration is that if the iron atom sits in the plane of the porphyrin, the proximal His F8 imidazole

Chains of Human Hemoglobin

Developmental Stage	*Symbol*	*Chain Designations*
Adult	HbA_1	$\alpha_2\beta_2$
Adult	HbA_2	$\alpha_2\delta_2$
Fetus	HbF	$\alpha_2\gamma_2$
Embryo	Hb Gower-1	$\zeta_2\varepsilon_2$
Embryo	Hb Portland	$\zeta_2\gamma_2$

will interact unfavourably with atoms of the porphyrin. The strength of this unfavourable steric interaction is due, in part, to conformational constraints on the His F8 and the porphyrin in the deoxyhemoglobin conformation that forces the approach of the His F8 toward the porphyrin to a particular path. These constraints become less significant in the oxy conformation of hemoglobin.

The conformation with the iron atom out of the plane of the porphyrin is unstrained and energetically favoured for the five-coordinate ferrous atom. When O_2 binds the sixth coordiante position of the iron, however, this conformation becomes strained. A more energetically favourable conformation for the O_2 liganded iron is one in which the iron atom is within the plane of the porphyrin structure.

On binding of O_2 to a ferrous atom the favourbale free energy of bond formation overcomes the repulsive interaction between His F8 and porphyrin, and the ferrous atom moves into the plane of the porphyrin ring. This is the most thermodynamically stable position for the now six-bonded iron atom; one axial ligand is on either side of the plane of the porphyrin ring, and the steric repulsion of one of the axial ligands with the porphyrin is balanced by the repulsion of the second axial ligand on the opposite side when the ferrous atom is in the centre. If the iron atom is displaced from the center, the steric interactions of the two axial ligands with the porphyrin in the deoxy conformation are unbalanced, and the stability of the unbalanced structure will be lower than that of the equidistant conformation. Also, the radius of the iron atom with six ligands is reduced so that it can just fit into the center of the porphyrin without distortion of the porphyrin conformation.

Since steric repulsion between porphyrin and His F8 in the deoxy conformation must be overcome on O_2 association, binding of the first O_2 is characterized by a relatively low affinity constant. However, when O_2 association occurs to the first heme in deoxyhemoglobin, the change in position of the iron atom from above the plane of the porphyrin into the center of the porphyrin triggers a conformational change in the whole molecule. The change in conformation results in a greater affinity of O_2 to the other heme sites after the first O_2 has bound.

The conformation of deoxyhemoglobin is stabilized by noncovalent interactions of the quaternary structure at the interface between α and β subunits in which the FG corner of one subunit noncovalently binds to the C helix of the adjacent subunit. In addition, ionic interactions stabilize the deoxy conformation of the protein. These interactions of the deoxy conformation are now

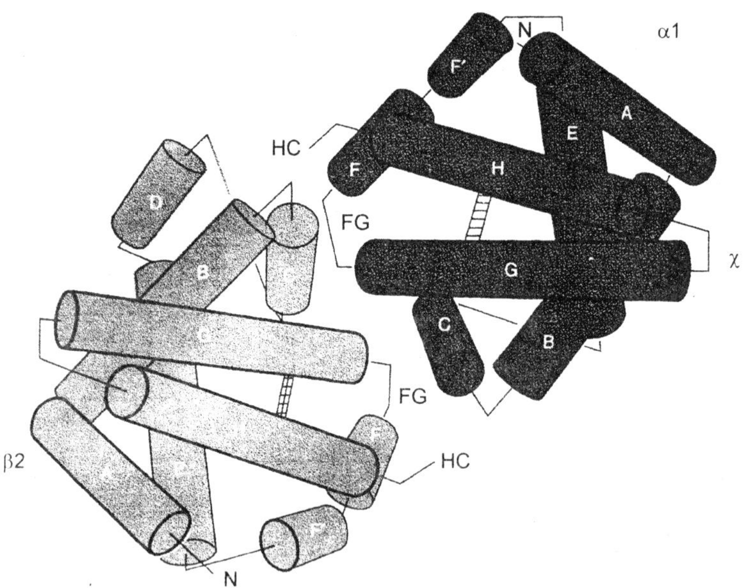

Quaternary structure of hemoglobin.

destabilized on the binding of O_2 to one of the heme subunits of a deoxyhemoglobin molecule. The binding of O_2 pulls the Fe^{2+} atom into the porphyrin plane and moves the His F8 toward the porphyrin and with it the F helix of which the His F8 is a part. Movement of the F helix, in turn, moves the FG corner of its subunit, destabilizing the FG noncovalent interaction with the C helix of the adjacent subunit at an $\alpha_1\beta_2$ or $\alpha_2\beta_2$ subunit interface.

The FG to C intersubunit contacts act as a "switch," because they exist in two different arrangements with different modes of contact between the FG corner of one subunit and the C. helix of the adjacent subunit. The switch in noncovalent interactions between the two positions involves a relative movement of FG and C in adjacent subunits of about 6 Å. In the second position of the "switch," the tertiary conformation of the subunits participating in the FG to C intersubunit contact is less constrained and the adjacent subunit changes to a new tertiary conformation (oxy conformation) even without O_2 bound.

This oxy conformation allows the His F8 residues to approach their porphyrins on O_2 association with a less significant steric repulsion than in the deoxy conformation. Thus an O_2 molecule binds to the empty hemes in the less constrained oxy conformation more easily than to a subunit conformation held by the quaternary interactions in the deoxy conformation.

In addition, Val E11 in the deoxy conformation of β subunits is at the entrance to the O_2-binding site, where it sterically impedes O_2 association to heme. In the oxy conformation the heme in β subunits moves approximately 1.5 Å further into the heme-binding site, changing the geometric relationship of the O_2-binding site to the Val E11 side chain, so that the Val E11 no longer sterically interferes with O_2 binding. This is an important additional factor that increases affinity of O_2 for the oxy conformation of the β chain over that for the deoxy conformation.

The deoxy conformation of hemoglobin is referred to as the "tense" or **T conformational state.** The oxyhemoglobin conformational form is referred to as the "relaxed" or **R conformational**

state. The allosteric mechanism shows how initial binding of the oxygen to one of the heme subunits of the tetrameric molecule pushes the molecular conformation from the T to R conformational state. The affinity constant of O_2 is greater for the R state hemes than the T state by a factor of 150-300, depending on the solution conditions.

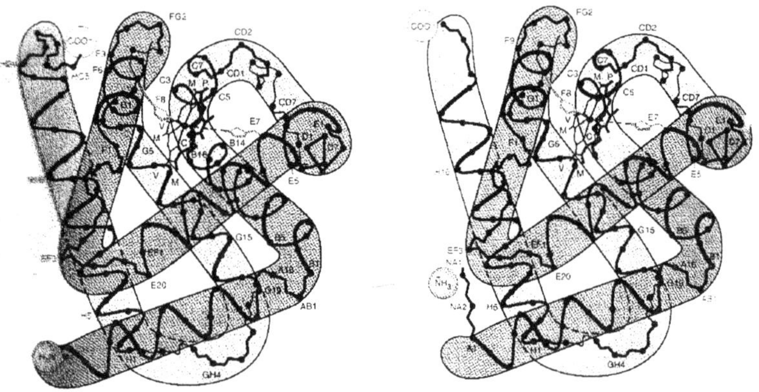

Comparison of conformation of (*a*) myoglobin and (*b*) β chain of HbA_1.

Biological Membranes

4.1 Composition of Membranes

Biological membranes from either eukaryotic or prokaryotic cells have the same classes of chemical components, a similarity in structural organization, and many properties in common. There are major differences in specific lipid, protein, and carbohydrate components but not in physicochemical interaction of these molecules. Membranes have a trilaminar appearance when viewed by electron microscopy with two dark bands on each side of a light band. The overall width of most mammalian membranes is 7-10 nm but some have significantly smaller widths. Intracellular membranes are usually thinner than plasma membranes. Many do not appear symmetrical, with an inner dense layer often thicker than an outer dense layer; there is a chemical asymmetry of membranes. With development of sophisticated techniques for preparation of tissue samples and staining, including negative staining and freeze fracturing, surfaces of membranes have been viewed; at the molecular level surfaces are not smooth but dotted with protruding globular-shaped masses.

Membranes are very dyamic structures with a movement that permits cells as well as subcellular structures to adjust their shapes and to change position. Chemical components of membranes, that is, lipids and protein, are ideally suited for their dynamic role. Membranes are an organized sea of lipid in a fluid state, a nonaqueous compartment of cells, in which various components are able to move and interact.

Cellular membranes control the composition of space that they enclose by excluding a variety of molecules and by selective transport systems allowing movement of specific molecules from one side to the other. These transporters are proteins. By controlling translocation of substrates, cofactors, ions, and so on, membranes modulate the concentration of substances in cellular compartments, thereby exerting an influence on metabolic pathways. Hormones, and growth and metabolic regulators bind to specific protein receptors on plasma membranes and the information to be imparted to the cell is transmitted by the membrane component to the appropriate metabolic pathway by a series of intracellular intermediates, termed second messengers. Plasma membranes of eukaryotic cells also have a role in cell-cell recognition, maintenance of the shape of cells, and cell locomotion.

The discussion that follows is directed to the chemistry and transport functions of membranes primarily of mammalian cells but the observations and activities described are applicable to all biological membranes.

Lipids and proteins are the two major components of all membranes. The amount of each varies greatly between different membranes. Protein ranges from about 20% in the myelin sheath to over 70% in the inner membrane of the mitochondria. Intracellular membranes have a high percentage of protein because of the large number of enzymic activities of these membranes. Membranes also contain a small amount of various polysaccharides in the form of glycoprotein and glycolipid; there is no free carbohydrate in membranes.

The three major lipid components of eukaryotic cell membranes are glycerophospholipids, sphingolipids, and cholesterol. Glycerophospholipids and sphingomyelin, a sphingolipid containing phosphate, are classified as **phospholipds.** Bacteria and blue-green algae contain glycerolipids where a carbohydrate is attached directly to the glycerol. Individual cellular membranes also contain small quantities of other lipids, such as triacylglycerol and diol derivatives.

Glycerophospholipids (phosphoglycerides) have a glycerol molecule as the basic component to which phosphoric acid is esterified at the α carbon and two long-chain fatty acids are esterified at the remaining carbon atoms. Glycerol does not contain an asymmetric carbon, but the α-carbon atoms are not stereochemically identical. Esterification of a phosphate to an α carbon makes the molecule asymmetric. The naturally occurring glycerophospholipids are designated by the stereospecific numbering system (*sn*).

1,2-Diacylglycerol 3-phosphate, **phosphatidic acid,** is the parent compound of a series of glycerophospholipids, where different hydroxyl-containing compounds are esterified to the phosphate. The major compounds attached by a phosphodiester bridge to glycerol are choline, ethanolamine, serine, glycerol, and inositol. **Phosphatidylethanolamine** (ethanolamine glycerophospholipids or the trivial name cephalin) and **phosphatidylcholine** (choline glycerophospholipid or lecithin) are the most common glycerophospholipids in membranes. **Phosphatidylglycerol phosphoglyceride** (diphosphatidylglycerol or cardiolipin) contains two phosphatidic acids linked by a glycerol and is found nearly exclusively in mitochondrial inner membranes and bacterial membranes.

Glycerol ether phospholipids contain a long aliphatic chain in ether linkage to the glycerol at the *sn*-1 position. Ether phospholipids contain an alkyl group (alkyl acylglycerophospholipid) or an α, β-unsaturated ether, termed a **plasmalogen.** The latter groups are more prevalent in membranes. Plasmalogens containing ethanolamine (ethanolamine plasmalogen) and choline (choline plasmalogen) esterified to the phosphate are abundant in nervous tissue and heart but not in liver. In human hearts more than 50% of the ethanolamine glycerophospholipids are plasmalogens.

The amino alcohols **sphingosine** (D-4-sphingenine) and **dihydrosphingosine** are the basis for another series of membrane lipids, the **sphingolipids.** A **ceramide** is sphingosine with a saturated or unsaturated long-chain fatty acyl group in amide linkage on the amino group. With two nonpolar tails a ceramide is similar in structure to diacylglycerol. Various substitutions are found on the hydroxyl group at position 1. The sphingomyelin series has phosphorylcholine esterified to the 1-OH and is the most abundant sphingolipid in mammalian tissues. The similarity of this structure to choline glycerophospholipids is apparent, and they have many properties in common; note that the sphingomyelins are amphipathic compounds with a charged head group. Sphingomyelins and glycerophospholipids are classified as phospholipids. The sphingomyelin of myelin contains predominantly the longer chain fatty acids, with carbon lengths of 24; as with glycerophospholipids, there is a specific fatty acid composition of the sphingomyelin, depending on the tissue.

Glycosphingolipids do not contain phosphate and have a sugar attached by a *β*-glycosidic linkage to the 1-OH group of the sphingosine in a ceramide. One subgroup is the **cerebrosides,** which contain either a glucose (**glucocerebrosides**) or galactose (**galactocerebrosides**) attached to a ceramide. Cerebrosides are neutral compounds. Galactocerebrosides are found predominantly in brain and nervous tissue, whereas the small quantities of cerebrosides in nonneural tissues usually contain glucose. **Phrenosin,** a specific galactocerebroside, contains a 2-OH C_{24} fatty acid. Galactocerebrosides may contain a sulfate group esterified on the 3 position of the sugar. They are called **sulfatides**. Cerebrosides and sulfatides usually contain fatty acids with 22-26 carbon atoms.

In place of monosaccharides, neutral glycosphingolipids often have 2 (dihexosides), 3 (trihexosides), or 4 (tetrahexosides) sugar residues attached to the 1-OH group of sphingosine. Diglucose, digalactose, *N*-acetylglucosamine, and *N*-acetyldigalactosamine are the usual sugars.

The most complex group of glycosphingolipids, the **gangliosides,** contain oligosaccharide head groups with one or more residues of sialic acid; these are amphipathic compounds with a negative charge at pH 7.0. The gangliosides represent 5-8% of the total lipids in brain, and some 20 different types have been identified differing in the number and relative position of the hexose and sialic acid residues. This is the basis of their classification.

4.2 Classification of Membrane Proteins

Depending on the basis of ease of removal from isolated membrane fractions, membrane proteins can be classified. They can be listed as follows.

1. Peripheral (extrinsic) proteins.
2. Integral (intrinsic)
3. Lipoproteins
4. Glycoproteins

Now let us learn one by one. Peripheral proteins are released from a membrane by treatment with salt solutions of different ionic strength or extremes of pH, and named to imply a physical location on the surface of the membrane. Peripheral proteins, many of which are enzymes, are usually soluble in water and free of lipids. **Integral (or intrinsic) proteins** require rather drastic treatment, such as use of detergents or organic solvents, to be separated from a membrane. They usually contain tightly bound lipid, which if removed leads to denaturation of the protein and loss of biological function. Integral proteins have sequences of hydrophobic amino acids, which create hydrophobic domains in the tertiary structure. These hydrophobic regions interact with the hydrophobic hydrocarbons of the lipids stabilising the protein-lipid complex. Removal of integral proteins leads to disruption of the membrane, whereas peripheral proteins can be removed with little or no change in the integrity of the membrane.

Proteolipids are hydrophobic **lipoproteins** soluble in chloroform and methanol but insoluble in water. They are present in many membranes but particularly in myelin, where they represent about 50% of the protein component. An example is lipophilin, a major lipoprotein of brain myelin that contains over 65% hydrophobic amino acids and covalently bound fatty acids.

Another class of integral membrane proteins is the **glycoproteins;** plasma membranes of cells contain a number of different glycoproteins, each with its own unique carbohydrate content.

The complexity, variety, and interaction of membrane proteins with lipids are just being resolved. Many of the proteins are enzymes located within or on the cellular membranes. Membrane proteins have a role in transmembrane movement of molecules and as receptors for the binding of hormones and growth factors. In many cells, such as neurons and erythrocytes, membrane proteins have a structural role to maintain the shape of the cell. Thus individual membrane proteins can have a *catalytic, transport, receptor, structural,* or *recognition role.* It is not surprising to find a high protein content in a membrane being correlated with the complexity and variety of functions of a membrane.

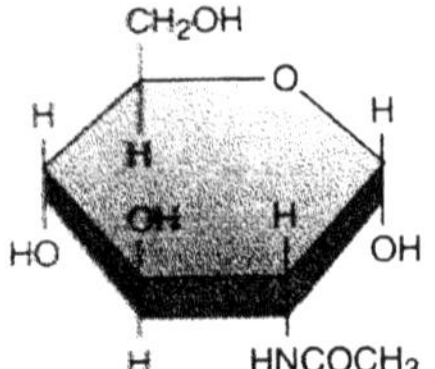

N-Acetyl-α-D-glucosamine

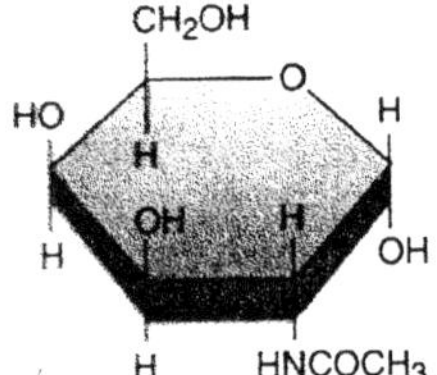

N-Acetyl-α-D-galactosamine

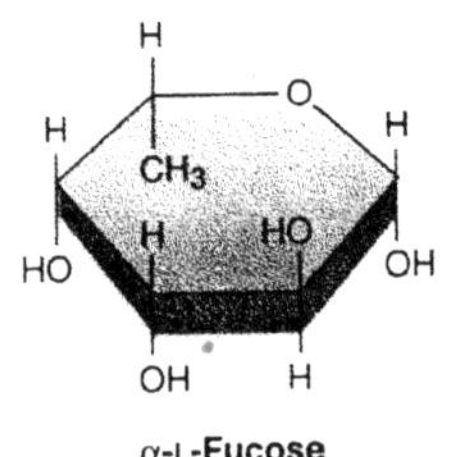

α-L-Fucose

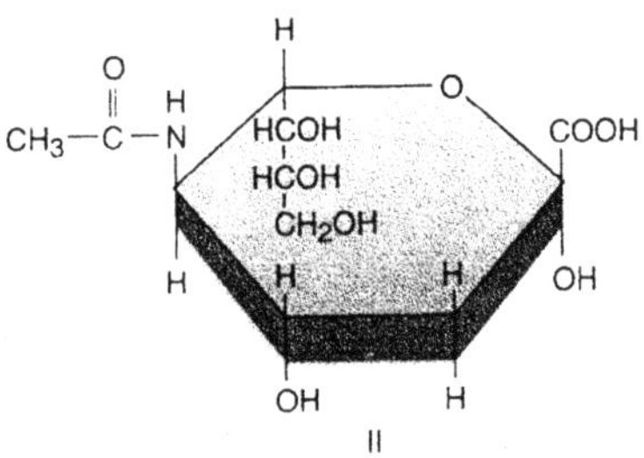

N-Acetyl-D-neuraminic acid

Structures of some membrane carbohydrates.

Carbohydrates present in membranes are **oligosaccharides** covalently attached to proteins to form glycoproteins and to a lessor amount to lipids to form glycolipids. The sugars found in glycoproteins and glycolipids include glucose, galactose, mannose, fucose, *N*-acetylgalactosamine, *N*-acetylglucosamine, and sialic acid. The carbohydrate is on the exterior side of the plasma membrane or the luminal side of the endoplasmic reticulum. Roles for membrane carbohydrates include cell-cell recognition, adhesion, and receptor action.

Artificial membrane systems have been studied extensively in order to understand the properties of biological membranes. A variety of techniques are available to prepare liposomes, using synthetic phospholipids and lipids extracted from natural membranes. Depending on the procedure, unilamellar and multilamellar (vesicles within vesicles) vesicles of various sizes (20 mm to 1 μm diameter) can be prepared. The interior of the vesicle is an aqueous environment, and it is possible to prepare liposomes with different substances entrapped. Both the external and internal environments of liposomes can be manipulated and properties—including ability to exclude molecules, interaction with various substances, and stability under different conditions—of these synthetic membranes have been studied. Na^+, K^+, Cl^-, and most polar molecules do not readily diffuse across lipid bilayers of liposomes, whereas the bilayer presents no barrier to water. Lipid-soluble nonpolar substances such as triacylglycerol and undissociated organic acids readily diffuse into the membrane remaining in the hydrophobic environment of the hydrocarbon chains. Proteins have been incorproated into liposomes to mimic a natural membrane. Membrane-bound enzymes and proteins involved in translocating ions have been isolated from various tissues and incorporated into the membrane of liposomes for evaluation of the protein's function. With liposomes it is easier to manipulate the various parameters of membrane systems and thus study various activities free of interfering reactions present in cell membranes. Liposomes are used in delivery of drugs in humans.

A major obstacle in the use of many drugs is lack of tissue specificity in the action of the drug. Administration of drugs orally or intravenously leads to a drug acting on many tissues and not exclusively on a target organ, resulting in toxic side effects. An example is the commonly observed suppression of bone marrow cells by anticancer drugs. Some drugs are metabolized rapidly and their period of effectiveness is relatively short. Liposomes have been prepared with drugs, enzymes, and DNA encapsulated inside and used as carriers for these substances to target organs. Liposomes prepared from purified phospholipids and cholesterol are nontoxic and biodegradable. Alteration of surface charge enhances drug incorporation and release. Attempts have been made to prepare liposomes for interaction at a specific target organ. Antibiotic, antineoplastic, antimalarial, antiviral, antifungal, and anti-inflammatory agents have been found to be effective when administered in liposomes. Some drugs have a longer period of effectiveness when administered encapsulated in liposomes. It may be possible to prepare liposomes with a high degree of tissue specificity so that drugs and perhaps even enzyme replacement can be carried out with this technique.

4.3 Membrane Structure

Based on evidence from physicochemical, biochemical, and electron microscopic investigations, knowledge of membrane structure has evolved. All biological membranes have a **bimolecular leaf** arrangement of lipids, as in liposomes. The amphipathic lipids and cholesterol are oriented so that hydrophobic portions of the molecules interact, minimizing their contract with water or other polar groups, and polar head groups of lipids are at the interface with the aqueous environment. J.D. Davson and J. Danielli in 1935 proposed this model for a membrane; their proposal was later refined by J.D. Robertson. A major question with the earlier models was how to explain the interaction of membrane proteins with the lipid bilayer. In the early 1970s. S.J. Singer and G.L. Nicolson proposed the mosaic model for membranes in which some proteins (intrinsic) are actually immersed in the lipid bilayer while others (extrinsic) are loosely attached to the surface of the membrane. It was suggested that some proteins spanned the lipid bilayer being in contact with the aqueous environment on both sides. The characteristics of the lipid bilayer explain many of the observed cellular membrane properties, including fluidity, flexibility that permits changes of shape and form, ability to self-seal, and impermeability. The model continues to undergo modification and refinement.

Integral membrane proteins have specific domains, for ligand binding, catalytic activity, and attachment of carbohydrate or lipid. The anion channel of the erythrocyte has two major domains: a hydrophilic amino-terminal domain on the cytosolic side of the membrane with binding sites for ankyrin, a protein that anchors the cytoskeleton and other cytosolic proteins, and a domain with 509 amino acids that traverses the membrane and mediates the exchange of Cl^- and HCO_3^-. Glycophorin contains 60% carbohydrate, all of which is attached to the protein domain on the extracellular side of the membrane. With such well-defined domains, integral membrane proteins have a defined orientation in the membrane rather than a random one. Specific structural orientation demonstrates another important aspect of membrane structure; biological membranes are asymmetric, with each surface having specific characteristics. The orientation of proteins is fixed during the synthesis of the membrane or replacement of the protein; the bulkiness of the proteins, as well as thermodynamic restrictions, prevents transverse (flip-flop) movement.

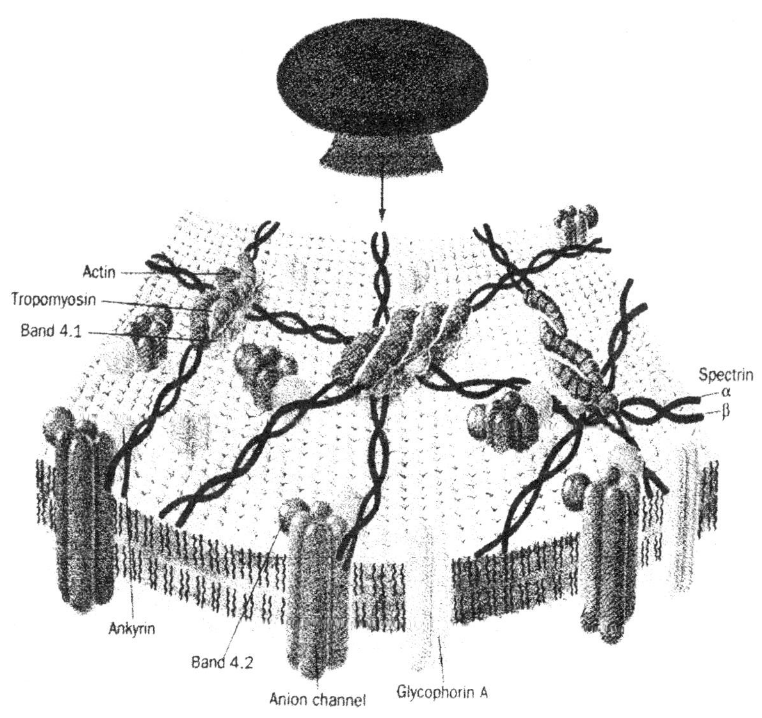

Schematic diagram of the erythrocyte membrane.

Many enzymes that are integral membrane proteins require the presence of the membrane lipid of activity. As an example, D-β-hydroxybutyrate dehydrogenase, located in the inner mitochondrial membrane, requires phosphatidylcholine for activity. Cholesterol has been implicated in the activity of various membrane ion pumps, including Na^+, K^+– and Ca^{2+}–ATPases, and acetylcholine receptors. Some of these modulating effects of lipids may be a reflection of a change in ordering and fluidity of the membrane but the lipid may also have a direct influence on the activity.

There is an **asymmetric distribution of lipid** components across biological membranes in contrast to the random distribution of lipids between the outer and inner lipid monolayers of liposomes. Each layer of the bilayer has a different composition with respect to individual glycerophospholipids and sphingolipids. An example is the asymmetric distribution of lipids in the human erythrocyte membrane. Sphingomyelin is predominantly in the outer layer, whereas phosphatidylethanolamine is predominantly in the inner lipid layer. In contrast, cholesterol is equally distributed on both sides of the plasma membrane.

Asymmetry of lipids may be maintained by specific membrane proteins that promote the transverse movement of specific lipids from one side to the other. Metabolic energy may be involved in this process. Uncatalyzed transverse movement from one side to the other (i.e., flip-flop movement) of the glycerophospholipids and sphingolipids is slow. The asymmetry of lipids in erythrocyte membranes is an example of how slow is the transverse movement of membrane lipids. Mature erythrocytes have a lifetime of about 120 days, during which there is no new membrane synthesis or even significant repair. Even so, there appears to be little mixing of phospholipids between molecular layers. Individual lipids can exchange with lipids in the cell matrix, as well as with lipids of other membranes. Specific mechanisms to maintian both the composition and asymmetry of lipids in membranes apparently eixst.

The specific composition of individual biological membranes leads to differences in fluidity. Glycerophospholipids containing short-chain fatty acids will increase the fluidity as does an increase in unsaturation of the fatty acyl groups. Cis double bonds in unsaturated fatty acids of phospholipids lead to kinks in the hydrocarbon chain, preventing the tight packing of the chains and creating pockets in the hydrophobic areas. It is assumed that these spaces, which will also be mobile due to the mobility of the hydrocarbon chains, are filled with water molecules and small ions. **Cholesterol** with its flat stiff ring structure reduces the coiling of the fatty acid chain and decreases fluidity. Consideration has been given to the potential clinical significance of high blood cholesterol on the fluidity of cell membranes. Ca^{2+} ion decreases the fluidity of membranes because of its interaction with the negatively charged phospholipids, reducing repulsion between polar groups and increasing packing of lipid molecules. This ion causes aggregation of lipids into clusters, reducing membrane fluidity.

Fluidity at different levels within the membrane also varies. The hydrocarbon chains of the lipids have a motion, which produces a fluidity in the hydrophobic core. The central area of a bilayer is occupied by ends of the hydrocarbon chains and is more fluid than areas closer to the two surfaces, where there are more constraints due to stiffer portions of the hydrocarbon chains. Cholesterol makes membranes more rigid toward the periphery because it does not reach into the central core of membrane.

Individual lipids and proteins move rapidly in a lateral motion along the surface of membranes. Electrostatic interactions of polar head groups, hydrophobic interactions of cholesterol with selected phospholipids or glycolipids, and protein–lipid interactions, however, lead to constraints on movement. There are lipid domains in which lipids move together as a unit.

Movement of integral membrane proteins in the lipid environment has been demonstrated by fusion of human and rat cells. When antigenic membrane proteins on cells of each species were labeled with different antibody markers, markers indicated the localisation of the proteins on the membrane. Immediately following fusion of the cells, proteins on the membranes of the human and rat cells were segregated in different hemispheres of the new cell, but within 40 minutes the two groups of proteins were evenly distributed over the membrane of the new cell. Movement of protein is slower than that of lipids and may be restricted by other membrane proteins, matrix proteins, or cellular structural elements such as microtubules or microfilaments to which they may be attached.

Membrane fluidity can control the activity of membrane-bound enzymes, membrane functions such as phagocytosis, and cell growth. A major factor in controlling the fluidity of the plasma membrane in higher organisms and mammals is the presence of cholesterol. With increasing cholesterol content the lipid bilayers become less fluid on their outer surface but more fluid in the hydrophobic core. Erythrocyte membranes of individuals with spur cell anemia have an increased cholesterol content. This condition occurs in severe liver disease such as cirrhosis of the liver in alcoholics. Erythrocytes have a spiny shape and are destroyed prematurely in the spleen. The cholesterol content is increased 25-65%, and the fluidity of the membrane is decreased. The erythrocyte membrane requires a high degree of fluidity for its function and any decrease would have serious effects on the cell's ability to pass through the capillaries.

4.4 Transport of Solute Molecules

Membranes of all cells contain highly specific transporters for the movement of inorganic anions and cations (e.g. Na^+, K^+, Ca^{2+}, HPO_4^{2+}, Cl^-, and HCO_3^-), and uncharged and charged organic compounds (e.g., amino acids and sugars). Different cellular membranes have different transport systems; as an example, the mitochondrial membrane has a specific mechanism to translocate ADP and ATP that is not present in other cellular membranes. Transport systems involve integral membrane proteins with a high degree of specificity for the substances transported. These proteins or protein complexes have been designated by a variety of names, including **transporter, translocase, translocator, permease,** and **pump,** or termed **transporter system, translocation mechanism,** and **mediated transport system.** The designations above are used interchangeably, but for convenience we will use transporter or translocase when referring to the proteins involved in translocation.

The characteristics of Membrane Transporters can be studied as follows.

Passive Mediated	*Active Mediated*
1. Saturation kinetics	1. Saturation kinetics
2. Specificity for solute transported	2. Specificity for solute transported
3. Can be inhibited	3. Can be inhibited
4. Solute moves down concentration	4. Solute can move against concentration gradient
5. No expenditure of energy	5. Requires coupled input of energy

If no energy is put into the system, the concentration on both sides of the membrane will be equal at equilibrium; but if there is an expenditure of energy, a concentration gradient can be established. Note the similarity of the role of a transporter to that of an enzyme; in both cases the protein increases the rate but does not determine the final equilibrium.

As the concentration of the substance to be translocated increases, the initial rate of transport increases but reaches a maximum when the substance saturates the protein transporter. Simple diffusion does not have saturation kinetics. Constants such as V_{max} and K_m can be calculated for transporters. As with enzymes, transporters can catalyze movement of a solute in both directions across the membrane depending on the $\Delta G'$ for the reaction.

Most transporters have a high degree of structural and stereo specificity for the substance transported. An example is mediated transport of D-glucose in erythrocytes, where the K_m for D-galactose is 10 times larger and for L-glucose 1000 times larger than for D-glucose. The transporter has essentially no activity with D-fructose or disaccharides. Competitive and noncompetitive inhibitors have been found for many transporters. Structural analogs of the substrate inhibit competitively and reagents that react with specific groups on proteins are noncompetitive inhibitors.

We need to expand the equation above and consider four aspects of mediated transport. These are: (1) *recognition by ransporter of appropriate solute from a variety of solutes in the aqueous environment,* (2) *translocation of solute across membrane,*

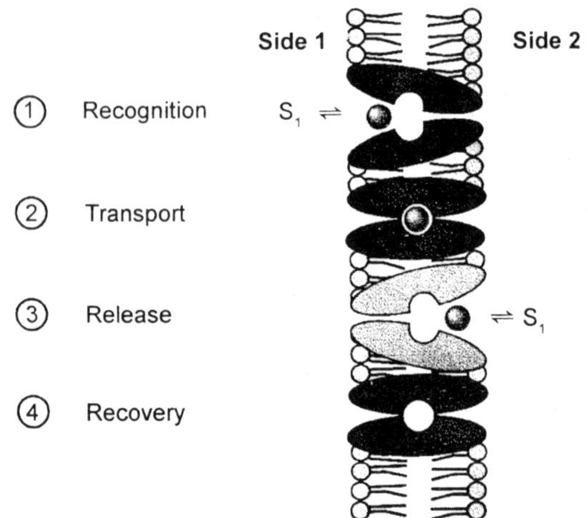

Model for a mediated transport system in a biological membrane.

(3) *release of solute by transporter,* and (4) *recovery of transporter to its original condition to accept another solute molecule.*

The first step, **recognition** of a specific substrate by the transporter, is explained on the same basis as that described for recognition of a substrate by an enzyme. The presence of very specific binding sites on the protein permits the transporter to recognize the correct structure of the solute to be translocated.

The second step, **translocation,** is not completely understood. A reasonable mechanism is one in which the protein transporter creates a channel between the environments on each side of the membrane with access through the channel being controlled by a gating mechanism in order to control which solutes can move into the channel. Transporters have receptor sites to which the solute attaches. After binding of solute and transporter, a conformational change of the protein moves the solute molecule a short distance, perhaps only 2 or 3 Å, but into the environment of the opposite side of the membrane. In this manner, it is not necessary for the transporter to move the molecule the entire distance across the membrane. Earlier suggestions for the translocation step included the possibility of a diffusible or rotating carrier, but both are improbable considering that transporters are large integral membrane proteins that do not diffuse transversely.

The movement of molecules by a transporter, with certain mechanism and direction can be noted as follows.

- Symport mechanism.

  (Two molecules move simultaneously in one direction)

- Antiport mechanism.

  (Two molecules move in opposite direction)

- Uniport mechanism.

  (Single molecule moves in one direction.)

If an oppositely charged ion is moved to balance the charge, the mechanism is called neutral or electrically silent.

4.5 Facilitated Diffusion

Facilitated Diffusion is generally referred as passive mediated transport, which leads to translocation of solutes through cell

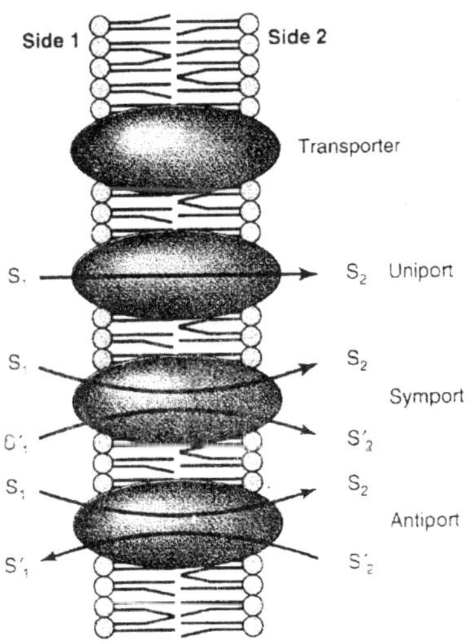

Uniport, symport, and antiport mechanisms for translocation of substances.

membranes without expenditure of metabolic energy. As with non-mediated diffusion the direction of flow is always from a higher to a lower concentration.

A family of passive mediated transporters for D-glucose, frequently referred to as **glucose permeases,** has been identified in the plasma membrane of mammalian cells. Six members have been described and are termed GLUT-1, GLUT-2, and so on. All have 12 hydrophobic segments considered to be the transmembrane regions. The physiological direction of movement is into the cell because the extracellular level of glucose is about 5 mM and most cells metabolize glucose rapidly, thus maintaining low intracellular concentrations. The transporter catalyzes a uniport mechanisms and is most active with D-glucose. D-Galactose, D-mannose, D-arabinose, and several other D-sugars as well as glycerol are translocated by the same transporter. L-Isomers are not transported. It has been proposed that the β-D-glucopyranose is transported with carrier interaction at the hydrogen atoms on at least C-1, C-3, and C-6 of the sugar. The affinity of erythrocyte translocase for D-glucose is highest with a K_m of ~6.2 mM, whereas for other sugars K_m values are much higher. The transporter has a very low affinity for D-fructose, precluding a role in cellular uptake of fructose; a separate carrier for fructose has been proposed. With isolated erythrocytes, glucose will move either into or out of the erythrocyte, depending on the direction of the experimentally established concentration gradient, demonstrating the reversibility of the system. Several sugar analogs as well as phoretin and 2,4,6-trihydro-xyacetophenone are competitive inhibitors.

An anion transporter in erythrocytes involves the antiport movement of Cl^- and HCO_3^-. The transporter is referred to as the **Cl^-–HCO_3^- exchanger, anion exchange protein,** or **band 3,** the latter because of its position in SDS polyacrylamide get electrophoresis of erythrocyte membrane proteins. The direction of ion flow is reversible and depends on the concentration gradients of the ions across the membrane.

The inner mitochondrial membrane contains several antiport systems for the exchange of anions between the cytosol and mitochondrial matrix. These include: (1) a transporter for exchange of ADP and ATP, (2) a transporter for exchange of phosphate and OH^-, (3) a dicarboxylate carrier that catalyzes an exchange of malate for phosphate, and (4) a translocator for exchange of aspartate and glutamate. In the absence of an input of energy these transporters will catalyze a passive exchange of metabolites down their concentration gradient to achieve a thermodynamic equilibrium of all intermediates. As an antiport mechanism, a concentration gradient of one compound can drive the movement of the other solute. In several cases, the transporter catalyzes the antiport movement of an equal number of charges on the substrate; in such movement the mitochondrial membrane potential influences the equilibrium and the anions can be moved against their concentration gradients. ADP-ATP and the phosphate transporters, as well as an uncoupling protein that translocates H^+, have significant amino acid homology and are presumably derived from a common ancestor. It has been suggested and each subunit has six transmembrane α-helices. The **uncoupling protein,** found in mitochondria of brown adipose tissue, has been proposed to be involved in generation of heat.

The **ATP-ADP translocase** is very specific for ATP and ADP and deoxyribose derivatives, dATP and dADP, but does not

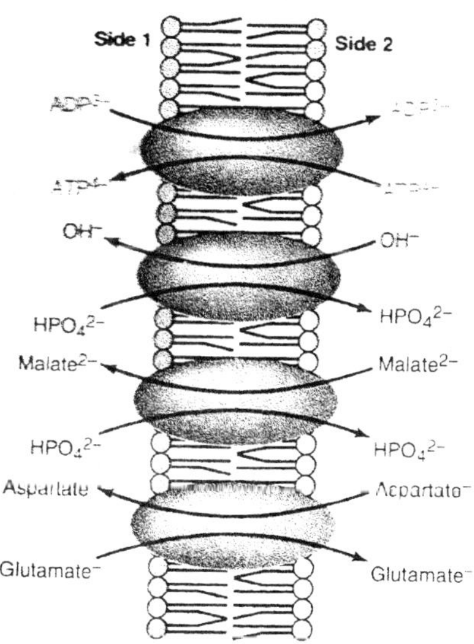

Representative anion transport systems in liver mitochondria.

transport AMP or other uncleotides. It is a dimer containing two subunits of 33 kDa each and represents about 12% of the total protein in heart mitochondria. It is very hydrophobic and can exist in two conformations. Atractyloside and bongkredic acid are specific inhibitor, each apparently reacting with a different conformation of the protein. The mitochondrial membrane potential can drive the movement of the nucleotides by this translocator, but in the absence of the membrane potential it functions as a passive mediated transporter.

It is sometimes difficult to differentiate passive mediated transport from simple diffusion but inhibition of the process is good evidence of a carrier.

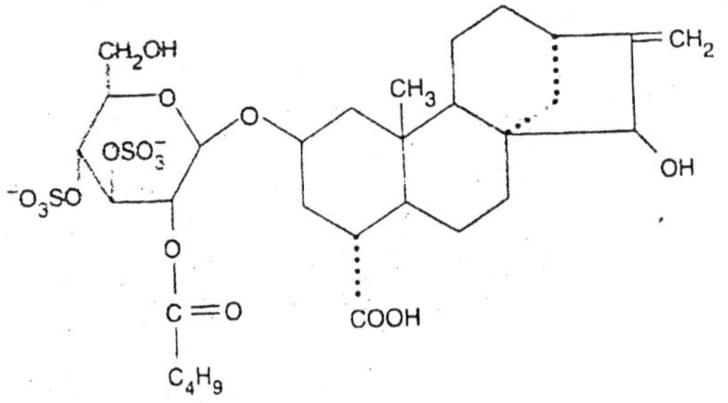

Atractyloside

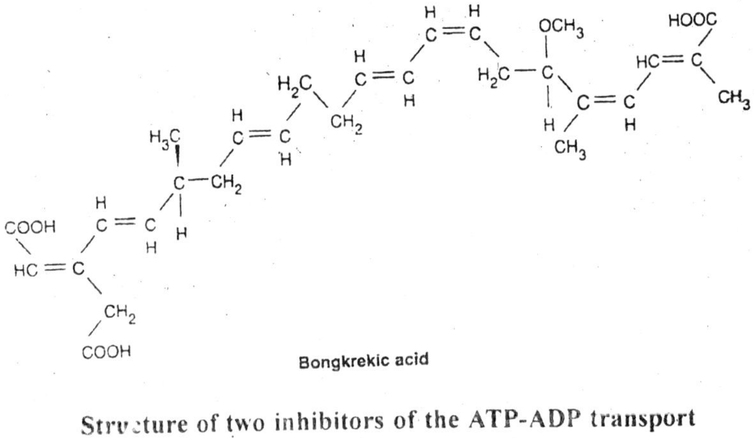

Bongkrekic acid

Structure of two inhibitors of the ATP-ADP transport system of liver mitochondria.

4.6 Transport System in Mammalian Cells

The transport systems are only indicative of the variety of transporters knows; others responsible for a variety of substances have been proposed. Most systems have been studied in only a few tissues and their localization may be more extensive than indicated.

The foregoing has presented the major mechanisms for movement of molecules across cellular membranes, particularly the plasma membrane. Cell organelles and membrane systems have a variety of transport systems. Mitochondria have transport mechanisms utilizing a proton gradient. Bacteria have transport systems analogous to those observed in mammalian cells.

Substance Transported	*Mechanism of Transport*	*Tissues*
Sugars		
Glucose	Passive	Widespread
	Active symport with Na^+	Small intestines and renal tubular cells
Fructose	Passive	Intestines and liver
Amino acids		
Amino acid-specific transporters	Active symport with Na^+	Intestines, kidney, and liver
All amino acids except proline	Active group translocation	Liver
Specific amino acids	Passive	Small intestine
Other organic molecules		
ATP-ADP	Antiport transport of nucleotides; can be active transport	Mitochondira
Ascorbic acid	Active symport with Na^+	Widespread
Biotin	Active symport with Na^+	Liver
Cholic acid, deoxycholic acid, and taurocholic acid	Active symport with Na^+	Intestines

(Contd...)

(Contd...)

Substance Transported	*Mechanism of Transport*	*Tissues*
Dicarboxylic acids	Active symport with Na^+	Kidney
Folate	Active	Widespread
Lactate and monocarboxylic acids	Active symport with H^+	Widespread
Neurotransmitters (e.g., γ-amino butyric acid, norepinephrine, glutamate, dopamine)	Active symport with Na^+	Brain
Organic anions (e.g., malate, α-ketoglutarate, glutamate)	Antiport with counter-organic anion	Mitochondria
Peptides (2 to 4 amino acids)	Active symport with H^+	Intestines
Urea	Passive	Erythrocytes and kidney
Inorganic ions		
H^+	Active	Mitochondria
H^+	Active, vacuolar ATPase	Widespread; lysosomes, endosomes, and Golgi complex
Na^+	Passive	Distal renal tubular cells
Na^+, H^+	Active antiport	Proximal renal tubular cells and small intestines
Na^+, K^+	Active: ATP driven	Plasma membrane of all cells
Na^+, HPO_4^{2-}	Active cotranspot	Kidney
Ca^{2+}	Active: ATP driven	Plasma membrane and endoplasmic (sarcoplasmic) reticulum
Ca^{2+}, Na^+	Active antiport	Widespread
H^+, K^+	Active antiport	Parietal cells of gastric mucosa secreting H^+
Cl^-/HCO_3^-	Passive antiport	Erythrocytes and other cells.

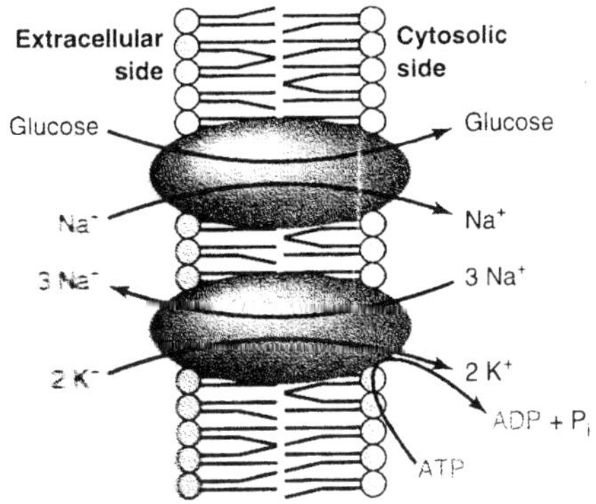

Na^+-dependent symport transport of glucose across the plasma membrane.

Enzymes

5.1 Introduction

The word **enzyme** was introduced by Kühne in 1878 to refer to the occurrence **in yeast** of something responsible for its fermentative activity. Berzelius, some 50 years earlier, had recognized that there were naturally occurring ferments which promoted chemical reactions and had forwarded the concept of biological catalysis. Berzelius classified the ferments into 'organized' and 'unorganized' on the basis of the presence or absence of intact micro-organisms. Kühne applied the word enzyme to ferments derived from extracts of, or secretions from, yeast, i.e. the 'unorganized ferments'. In 1897, Büchner prepared by filtration of the yeast extracts the first **cell-free** enzymic extract which could catalyse fermentation.

The **chemical nature** of enzymes was controversial. In 1926, Sumner crystallized **urease** from jack bean extracts but the preparation had so little activity that other workers attributed the catalytic effect to an active contaminant rather than the protein. Willstätter, on the other hand, purified **peroxidase** with such high activity that catalysis was demonstrated in preparations which were considered protein-free. During the period from 1930 to 1936, numerous enzymes including pepsin, trypsin and chymotrypsin were crystallized and analysed leading to the acceptance that enzymes are proteins.

Enzymes are specialized proteins that function in the acceleration of chemical reactions. Many reactions required for normal activity of cells would not proceed fast enough at the pH and temperature of the body without these specialized proteins. The term defining the speed of a chemical reaction, whether catalyzed or uncatalyzed, is **rate** or **velocity.** Rate (velocity) is the change in amount (moles, grams) of starting materials or products per unit time. Enzymes increase the rate by acting as catalysts. A **catalyst** increases the rate of a chemical reaction but is not itself changed in the process. An enzyme may become temporarily covalently bound to a molecule being transformed during intermediate stages of the reaction but at the end of the reaction the enzyme will again be in its original form as the product is released.

Two important characteristic of enzyme catalysts are that the enzyme is not changed as a result of catalysis and the enzyme does not change the equilibrium constant of the reaction but simply increases the rate at which the reaction approaches equilibrium. As will be discussed later, it accomplishes the rate increase by lowering the barrier to reaction; that is, it lowers the energy of activation. Therefore a catalyst increases the rate but does not change the thermodynamic properties of the system with which it is interacting.

Several terms need to be defined before we enter into a discussion of the mechanism of enzyme action. An **apoenzyme** is the protein part of an enzyme without any cofactors or prosthetic groups that may be required for the enzyme to be functional. The apoenzyme is catalytically inactive. Not all enzymes require cofactors or prosthetic groups. **Cofactors** are small organic or inorganic molecules that an apoenzyme requires for its activity. For example, in lysine oxidase copper is loosely bound but is required for the enzyme to be active. A **prosthetic,** group is similar to a cofactor but is tightly bound to an apoenzyme. For example, in the cytochromes, the heme prosthetic group is very tightly bound and requires strong acids to disassociate it from the apocytochrome. Addition of a cofactor or prosthetic group to the apoprotein yields the **holoenzyme,** that is, the active enzyme. The molecule acted upon by the enzyme to form product is the **substrate.** Since most reactions are reversible, the products of the forward reaction become substrates for the reverse reaction.

Enzymes have a great deal of specificity. For example, glucose oxidase will oxidize glucose but not galactose. The specificity resides in a particular region on the enzyme surface, the **substrate-binding site,** a particular arrangement of amino acid side chains in the polypeptide that is specially formulated to bind a specific substrate.

Some enzymes have broad specificity; glucose, manose, and fructose are phosphorylated by hexokinase, whereas glucokinase is specific for glucose. The substrate-binding site may contain the **active site.** In some cases, however, the active site may not be within the substrate-binding site but may be contiguous to it in the primary sequence. In other instances the active-site residues lie in distant regions of the primary sequence but are brought adjacent to the substrate-binding site by folding in the tertiary structure. The **active site** contains the machinery, in the form of particular amino acid side chains, involved in catalyzing the reaction.

Some enzymes have variants called **isoenzymes** (isozymes) that catalyze the same chemical reaction. Isoenzymes are electrophoretically distinguishable because of mutations in one or more amino acids in noncritical areas of the protein.

Some enzymes have a region of the molecule, the **allosteric** site, that is not at the active site or substrate-binding site but is a unique site where small molecules bind and effect a change in the substrate-binding site or the activity occurring in the active site. The binding of a specific small molecule at the allosteric site causes a change in the conformation of the enzyme. This can cause the active site to become either more active or less active by increasing or decreasing the affinity of the binding site for substrate.

5.2 Classification of Enzymes

By the late 1950s, enzyme nomenclature was in confusion. Without a guiding authority, the increase in known enzymes had led to the assignment of misleading or inappropriate names. In many cases the same enzyme had become known by several names, while some catalytically different enzymes had been identically named. In 1961, the first Enzyme Commission reported a system for enzyme classification and the assignment of code numbers.

The International Union of Biochemistry and Molecular Biology (IUBMB) has established a system whereby all enzymes are classified into six major classes, each subdivided into subclasses that are further subdivided. In naming an enzyme, the substrates are stated first, followed by the reaction type to which the ending -ase is affixed. For example, alcohol dehydrogenase is alcohol: NAD^+ oxidoreductase because it catalyzes an oxidation-reduction reaction and the electron donor is an alcohol and the acceptor is NAD^+. Many common names persist but are not very informative. For example, "aldolase" does not tell much about the substrates, although it does identify the reaction type. We will use trivial names recognized by the IUMBM and that are in common usage.

Now let us study about the six major classes of enzymes.

Class (i) Oxidoreductoses

The enzymes catalyze **oxidation-reduction reactions.** For example, alcohol:NAD^+ oxidoreductase (alcohol **dehydrogenase**) catalyzes the oxidation of an alcohol to an aldehyde. It removes two electrons and two hydrogen atoms from the alcohol to yield an aldehyde, and, in the process, the two electrons originally in the carbon-hydrogen bond of the alcohol are transferred to the NAD^+, which is reduced. NAD^+, is a cofactor that mediates many biological oxidation-reduction reactions. In addition to the alcohol and aldehyde functional groups, dehydrogenases also act on the following functional groups as electron donors: $—CH_2—CH_2—$, $—CH_2—NH_2$, and —CH=NH, as well as the cofactors NADH and NADPH.

There are other subclasses of the oxidoreductases, **Oxidases** transfer two electrons from the donor to oxygen, resulting usually in hydrogen peroxide (H_2O_2) formation. **Cytochrome oxidase** produces H_2O rather than H_2O_2. **Oxygenases** catalyze the incorporation of oxygen into a substrate. With dioxygenases both atoms of O_2 are incorporated in a single product.

Class (2) Transferases

These enzymes transfer functional groups between donors and acceptors. The amino, acyl, phosphate, one-carbon, and glycosyl groups are the major moieties that are transferred. **Amino-**

transferases (transaminases) transfer an amino group from one amino acid to an α-keto acid acceptor, resulting in the formation of a new amino acid and a new keto acid. **Kinases** are the phosphorylating enzymes that catalyze the transfer of the γ phosphoryl groups from ATP or another nucleoside triphosphate to alcohol or amino group acceptors. For example, glucokinase catalyzes the phosphorylation of glucose.

Glycogen synthesis depends on **glucosyltransferases,** which transfer an activated glucosyl residue to a glycogen primer. The phosphoester bond in uridine diphosphoglucose is labile, which allows the glucose to be transferred to the growing end of the glycogen primer.

Although a polymer is synthesized, the reaction is not of the ligase type reaction.

Class (3) Hydrolases

This group of enzymes can be considered as a special class of the transferases in which the donor group is transferred to water. The generalized reaction involves the hydrolytic cleavage of C—O, C—N, O—P, and C—S bonds. The cleavage of a peptide bond is a good example of this reaction:

$$R_1-\overset{\overset{\displaystyle O}{\|}}{C}-NH-R_2+H_2O \rightarrow R_1-\overset{\overset{\displaystyle O}{\|}}{C}-O^- + H_3\overset{+}{N}-R_2$$

Proteolytic enzymes are a special class of hydrolases called **peptidases.**

Class (4) Lyases

Lyases add or remove the elements of water, ammonia, or carbon dioxide. **Decarboxylases** remove the element of CO_2 from α- or β-keto acids or amino acids:

$$R-\overset{\overset{\displaystyle O}{\|}}{C}-\overset{\overset{\displaystyle O}{\|}}{C}-O^- + H^+ \rightarrow R-\overset{\overset{\displaystyle O}{\|}}{C}-H + CO_2$$

Dehydratases remove H_2O in a dehydration reaction. Fumarase converts fumarate to malate.

Class (5) Isomerases

This very heterogeneous group of enzymes catalyze isomerizations of several types. These include cis-trans and aldose-ketose interconversions. Isomerases that catalyze inversion at asymmetric carbon atoms are either **epimerases** or **racemases.** **Mutases** involve intramolecular transfer of a group such as a phosphoryl. The transfer may be direct but can involve a phosphorylated enzyme as an intermediate. Phosphoglycerate mutase catalyzes conversion of 2-phosphoglycerate to 3-phosphoglycerate.

Class (6): Ligases

Since to ligate means to bind, these enzymes are involved in synthetic reactions where two molecules are joined at the expense of a "high-energy phosphate bond" of ATP. The term **synthetase** is reserved for this particular group of enzymes. The formation of aminoacyl tRNAs, acyl coenzyme A, and glutamine and the addition of CO_2 to pyruvate are reactions catalyzed by ligases. Pyruvate carboxylase is a good example of a ligase enzyme. The substrates bicarbonate and pyruvate are ligated to form a four-carbon (C4) α-keto acid.

Enzymes have been employed as **therapeutic agents** in the management of a variety of diseases. **Proteases, amylases** and **lipases** have long been used in preparations to aid digestion.

Asparaginase has been employed in the treatment of **acute lymphocytic leukaemia** (ALL). The rationale is that normal cells are capable of synthesizing asparagine whereas certain tumour cells including those in ALL are not able to synthesize this amino acyl derivative. Intravenous administration of asparaginase destroys blood asparaginase and thereby lowers the concentration available to the tumour, thus reducing its growth.

Deoxyribon''clease (DNase) is employed in the treatment of **cystic fibrosis.** major feature of cystic fibrosis is recurrent respiratory infections, particularly due to *Pseudomonas aeroginosa,* which results in the release of large quantities of DNA from destroyed bacteria and dead phagocytic cells of the immune system. The DNA contributes to the viscosity of the mucus which causes some of the distressing symptoms. DNase may be used to degrade the DNA and thin the mucus so that the mucus can be dislodged by the patient's ciliary action.

5.3 Enzyme Activity and Substrate

Enzyme activity is usually expressed as micromoles (μmol) of substrate converted to product per minute under specified assay conditions. One standard unit of enzyme activity (U) is that activity that catalyzes transformation of 1 μmol min^{-1}. **Specific activity** of an enzyme preparation is defined as the number of enzyme units per milligram of protein (μmol min^{-1} mg of protein^{-1} or U/mg of protein). This expression, however, does not indicate whether the sample tested contains only the enzyme protein; during enzyme purification the value will increase as contaminating protein is removed. The **catalytic constant,** or **turnover number,** of an enzyme is equal to the units of activity per mole of enzyme (μmol/min/mol of enzyme). Where the enzyme has more than one catalytic center, the catalytic constant is often given on the basis of the particle weight of the subunit rather than the molecular weight of the entire protein. The Commission on Enzyme Nomenclature of the International Union of Biochemistry and Molecular Biology recommends that enzyme activity be expressed in units of moles per second, instead of micromoles per minute, to conform with the rate constants used in chemical kinetics. A new unit, the **katal** (abbreviated kat), is proposed where 1 kat denotes conversion of 1 mol substrate per second. Activity can be expressed, however, as millikatals (mkat), microkatals (μkat), and so forth. The specific activity and catalytic constant can also be expressed in katals.

The catalytic constant or turnover number allows direct comparison of relative catalytic ability between enzymes. For example, the constants for catalase and α-amylase are 5×10^6 and 1.9×10^4, respectively, indicating that catalase is about 2500 times more active than amylase.

Maximum velocity, V_{max}, is the velocity obtained under conditions of substrate saturation of the enzyme under specified conditions of pH, temperature, and ionic strength. V_{max} is a constant for a given enzyme.

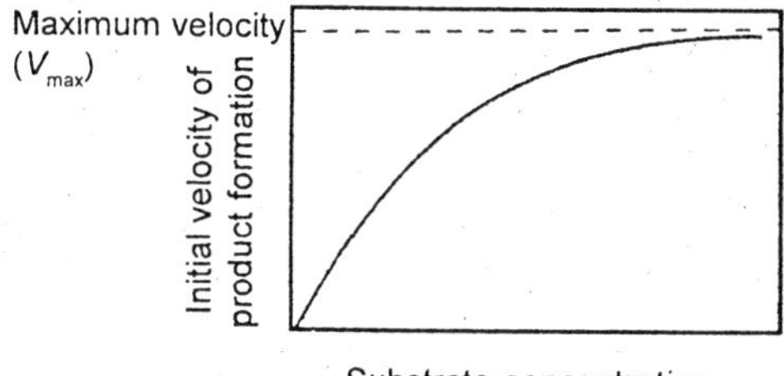

Plot of velocity versus substrate for an enzyme-catalyzed reaction.

The initial velocity of an enzyme-catalyzed reaction is dependent on the concentration of substrate (S). As concentration of substrate increases (S_1–S_4), initial velocity increases until the enzyme is completely saturated with substrate. If initial velocities obtained at given substrate concentrations are plotted, a rectangular hyperbola is obtained like that obtained for binding of oxygen to myoglobin as a function of increasing oxygen pressure. In general, a rectangular hyperbola is obtained for any process that involves interaction or binding of reactants or other substances at a specific but limited number of sites. The velocity of the reaction reaches a maximum at the point at which all the available sites are saturated.

The curve is referred to as the substrate saturation curve of an enzyme-catalyzed reaction and reflects the fact that an enzyme has a specific binding site for the substrate. Enzyme (E) and substrate must interact in some way if the substrate is to be converted to products. Initially there is formation of a complex between the enzyme and substrate:

$$E + S \underset{k_2}{\overset{k_1}{\rightleftharpoons}} ES$$

The rate constant for formation of the ES complex is defined as k_1, and the rate constant for disassociation of the ES complex is defined as k_2. So far, we have described only an equilibrium binding of enzyme and substrate. The chemical event in which bonds are made or broken occurs in the ES complex. The conversion of substrate to products (P) then occurs from the ES complex with a rate constant k_3. Therefore, transformed to

$$E + S \underset{k_2}{\overset{k_1}{\rightleftharpoons}} ES \xrightarrow{k_3} E + P$$

This equation is a general statement of the mechanism of enzyme action. The equilibrium between E and S can be expressed as an affinity constant, K_a, only if the rate of the chemical phase of the reaction, k_3, is small compared to k_2; then $K_a = k_1/k_2$. Earlier we used K_{eq} to describe chemical reactions. In enzymology the association or affinity constant K_a is preferred.

The **initial velocity,** v_0 of an enzyme-catalyzed reaction is dependent on amount of substrate present and on enzyme concentration. There is enough substrate initially to saturate the enzyme at all levels. The initial velocity doubles as the concentration of enzyme doubles. At the lower concentrations of enzyme, equilibrium is reached more slowly than at higher concentrations, but the final equilibrium position is the same.

From this discussion, we can conclude that the velocity of an enzyme reaction is dependent on both substrate and enzyme concentrations.

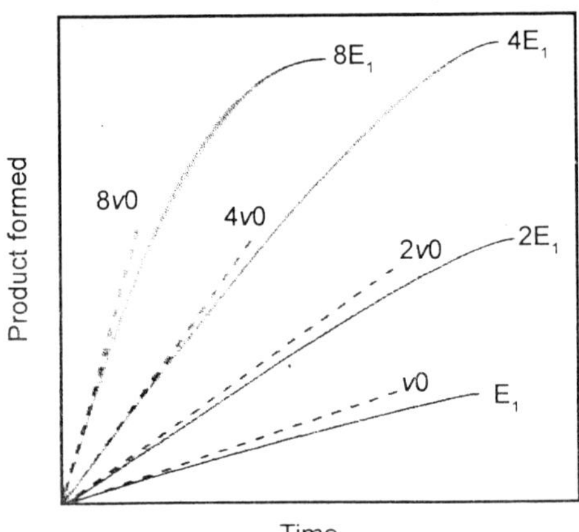

Progress curves at variable concentrations of enzyme and saturating levels of substrate.

5.4 Coenzymes and their Function

Coenzymes are small organic molecules, often derivatives of vitamins, that function with the enzyme in the catalytic process. Often the coenzyme has an affinity for the enzyme that is similar to that of the substrate; consequently, the coenzyme can be considered to be a second substrate. In some cases, the coenzyme is covalently bound to the apoenzyme and functions at or near the active site in catalysis. In other enzymes the role of the coenzyme falls between these two extremes.

Several coenzymes are derived from the B vitamins. Vitamin B_6, pyridoxine, requires little modification to form the active coenzyme, pyridoxal phosphate. Clinical correlation 4.3 points out the importance of the coenzyme-binding site and how alterations in this site cause metabolic dysfunction.

In contrast to vitamin B_6, niacin requires major alteration in mammalian cells to form a coenzyme.

The structures and functions of the coenzymes of only two B vitamins, niacin and riboflavin, and of ATP are discussed in this chapter. The structures and functions of coenzyme A (CoA), thiamine, biotin, and vitamin B_{12} are included in those chapters dealing with enzymes dependent on the given coenzyme for activity.

Adenosine triphosphate (ATP) often functions as a second substrate but can also serve as a cofactor in modulation of the activity of specific enzymes. This compound is central in biochemistry and it is synthesized *de novo* in all mammalian cells. The nitrogenous heterocyclic ring is adenine. The combination of the base, adenine, plus ribose is known as adenosine; hence ATP is adenosine that has at the 5′-hydroxyl a triphosphate. The biochemically functional end is the reactive triphosphate. The terminal phosphate-oxygen bond has a high free energy of hydrolysis, which means that the phosphate can be transferred from ATP to other acceptor groups. For example, as a cosubstrate ATP is utilized by the kinases for the transfer of the terminal phosphate to various acceptors. A typical example is the reaction catalyzed by **glucokinase:**

$$\text{Glucose} + \text{ATP} \rightarrow \text{glucose 6-phosphate} + \text{ADP}$$

ADP is adenosine diphosphate.

ATP also serves as a modulator of the activity of some enzymes. These enzymes have binding sites for ATP, occupancy of which changes the affinity or reactivity of the enzyme toward its substrates. In these cases, ATP acts as an **allosteric effector**.

Niacin is pyridine-3-carboxylic acid. It is converted to two coenzymes involved in oxidoreductase reactions. They are **NAD (nicotinamide adenine dinucleotide)** and **NADP (nicotinamide adenine dinucleotide phosphate).** The abbreviations NAD and NADP are convenient to use when referring to the coenzymes regardless of their state of oxidation or reduction. NAD^+ and $NADP^+$ represent the oxidized forms, and NADH and NADPH represent the reduced forms. Some dehydrogenases are specific for NADP and others for NAD; some function with either coenzyme. This arrangement allows for specificity and control over dehydrogenases that reside in the same subcellular compartment.

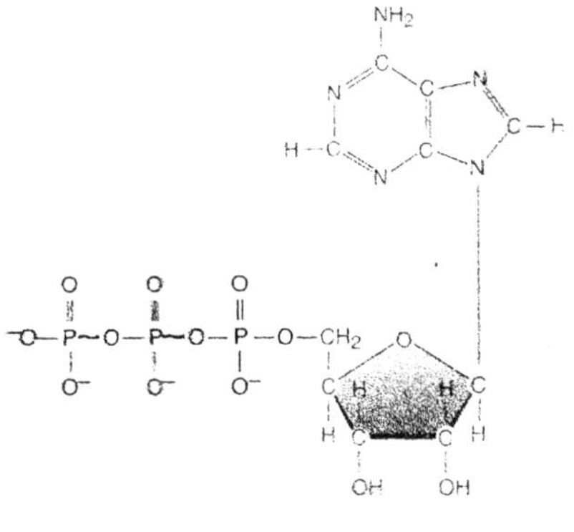

Adenosine triphosphate (ATP).

NAD^+ consists of adenosine and *N*-ribosyl-nicotinamide linked by a pyrophosphate linkage between the 5′-OH groups of the two ribosyl moieties. NADP differs structurally from NAD in having an additional phosphate esterified to the 2′-OH group of the adenosine moiety. Both coenzymes function as intermediates in transfer of two electrons between an electron donor and an acceptor. The donor and acceptor need not be involved in the same metabolic pathway. Thus the reduced form of these nucleotides acts as a common "pool" of electrons that arise from many oxidative reactions and can be used for various reductive reactions.

The adenine, ribose, and pyrophosphate components of NAD are involved in binding of NAD to the enzyme. Enzymes requiring NADP do not have a conserved aspartate residue present in the NAD-binding site. If the aspartate were present, a charge-charge interaction between the negative charged aspartate and the 2′-phosphate of NADP would prevent binding. Since there is no negatively charged phosphate on the 2′-OH in NAD, there is discrimination between NAD and NADP binding. The nicotinamide reversibly accepts and donates two electrons at a time. It is the active center of the coenzyme. In oxidation of deuterated ethanol by alcohol dehydrogenase, NAD^+ accepts two electrons and the deuterium from the ethanol. The other hydrogen is released as a H^+.

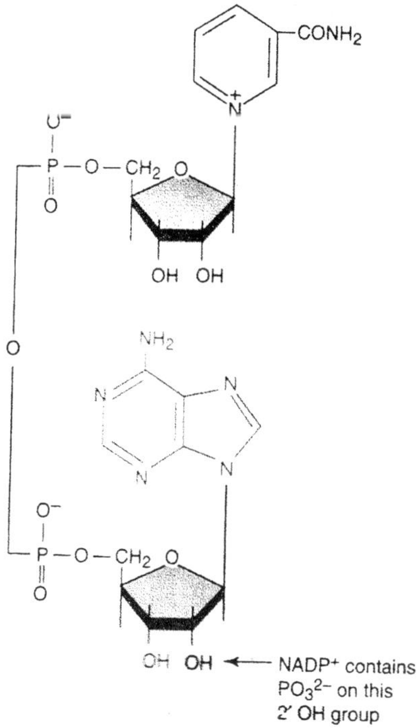

Nicotinamide adenine dinucleotide (NAD^+).

The binding of NAD^+ to the enzyme surface confers a chemically recognizable "top side" and "bottom side" to the planar nicotinamide. The former is known as the **A** face and the latter as the **B** face. In the case of alcohol dehydrogenase, the proton or deuterium ion that serves as a tracer is added to the **A** face. Other dehydrogenases utilize the **B** face. This particular effect demonstrates how enzymes can induce **stereospecificity** in chemical reactions by virtue of the asymmetric binding of coenzymes and substrates.

The two coenzyme forms of riboflavin are **FMN (flavin mononucleotide)** and **FAD (flavin adenine dinucleotide).** The vitamin riboflavin consists of the heterocyclic ring, isoalloxazine (flavin) connected through N-10 to the alcohol ribitol FMN has a phosphate esterfied to the 5′-OH group or ribitol. FAD is structurally analogous to NAD in having adenosine linked by a pyrophosphate linkage to a heterocyclic ring, in this case riboflavin. Both FAD and FMN function in oxidoreduction reactions by accepting and donating 2e$^-$ in the isoalloxazine ring. A typical example of FAD participation in an enzyme reaction is the oxidation of succinate to fumarate by succinate

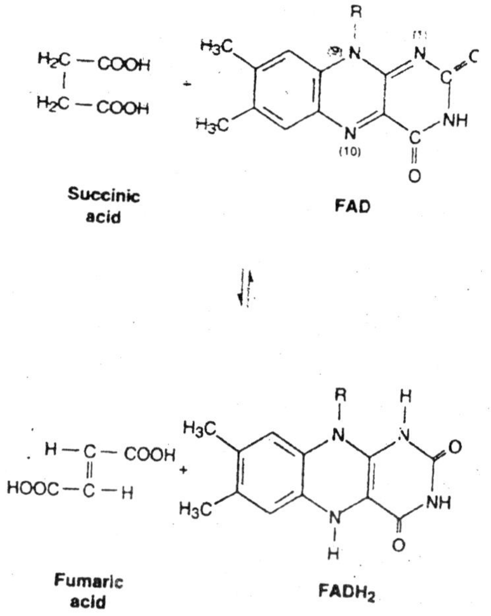

$H_2C-COOH$ / $H_2C-COOH$ + FAD

Succinic acid FAD

⇅

$H-C-COOH$ / $HOOC-C-H$ + $FADH_2$

Fumaric acid $FADH_2$

FAD as a coenzyme in the succinic dehydrogenase reaction.

5.5 Reversible and Irreversible Inhibition

Reversible inhibition is characterized by an equilibrium between free and inhibitor-bound enzymic forms. Stready-state kinetics may therefore be applied to the analysis of these inhibitions. The inhibitor may be removed from the enzyme-inhibitor (EI) complex by simple methods, e.g. dialysis, to deliver active enzymes. Various forms of reversible inhibition have been identified through kinetic studies:

- **competitive inhibition** in which the inhibitors have an effect on the K_m but not on the V of an enzyme;
- **non-competitive inhibition** which is characterized by a change in the V but not on the K_m of an enzyme;
- **uncompetitive inhibition** which is characterized by equal effects on both V and K_m of an enzyme.

Competitive inhibitors often bear a structural similarity to the substrate and compete with the substrate for the active sites of the enzyme, e.g. malonate inhibition of succinate dehydrogenase. However, competitive inhibitors are not necessarily structurally analogous to the substrate, e.g., salicylate inhibition of 3-phosphoglycerate kinase.

In competitive inhibition, the V is unchanged because the number of functional active sites is not altered but a greater substrate concentration is required to achieve the maximum utilization of the sites. Consequently, the K_m for the substrate increases. Competitive inhibition may therefore be overcome by the addition of more substrate to the enzyme reaction mixture.

Non-competitive inhibition is not reversed by additions of substrate. This type of inhibition occurs when an inhibitor binds to a site other than the active site on the free enzyme or ES complex. Non-competitive inhibitors interfere with either the formation of ES complex or its breakdown to yield product. Examples include heavy metal ions, e.g. Hg^{2+} and Pb^{2+}, which bind to strategically positioned sulphydryl groups and modulate the conformation of the enzyme. This mechanism is more common in reactions involving more than one substrate.

Uncompetitive inhibitors bind only to a formed ES complex. This inhibition mechanism is rare in one-substrate reactions but may occur as a type of product inhibition in reactions with multiple substrates and products.

Irreversible enzyme inhibitors are usually toxic substances. Irreversible inhibitors may be natural or designed synthetic substances.

The organophosphate **diisopropylfluorophosphate (DFP)** was designed to inhibit serine proteases by interaction with the hydroxyl group of the catalytic seryl residue to form a covalent adduct. DFP therefore inhibits irreversibly any enzyme whose catalytic site contains this essential seryl residue, e.g. acetylcholinesterase.

Acetylcholinesterase is crucial for nervous conduction, its inhibition resulting in the loss of motor or sensory function. Numerous insecticides and nerve gases, e.g. sarin, were based upon DFP. Sarin was employed in the terrorist attack on Tokyo underground railway system in March 1995 which resulted in ten deaths.

DFP and sarin are examples of **transition state analogues,** i.e. molecules which resemble the transition state. Other irreversible inhibitors closely resemble the substrate, e.g. **TPCK** was designed as an affinity label for chymotrypsin to allow the phenyl group to fit into a pocket in the enzyme's catalytic site.

Among the natural irreversible inhibitors are the **penicillin antibiotics** which inhibit a serine-containing enzyme responsible for linking the pentaglycine bridge to the tetrapeptide side chain during the biosynthesis of the bacterial cell wall.

Irreversible Inhibition

Irreversible inhibitors bind covalently to some functional group of the enzyme and permanently prevent their activity. The substrate is either hindered from entering the active site or is not converted into product and leaves the active site unchanged. In this type of inhibition, the effective concentration of the enzyme declines progressively as irreversible covalent bonding occurs. For this reason, irreversible inhibition cannot be analysed by Michaelis-

Menten kinetics. This type of inhibition is frequently used to obtain information regarding the functional amino acids at the active site, e.g. in affinity labels.

Suicide inhibition is a form of irreversible inhibition in which the substrate in the first catalytic cycle is converted into a chemically reactive product which remains covalently bound to the active site, thereby rendering the enzyme permanently inactive. Such inhibitors have potential as designed therapeutic agents with minimal side-effects. For example, **difluoromethylornithine** (DFMO) is a suicide inhibitor of **ornithine decarboxylase.** In mammalian cells, ornithine decarboxylase is rapidly degraded and replaced by continuous synthesis of the enzyme. In trypanosomes, this enzyme is stable and is not subjected to rapid turnover. DFMO has proved effective against the parasitic disease, trypanosomiasis (African sleeping sickness).

When covalent modification occurs at the binding site or the active site, inhibition will not be reversed by dialysis unless the linkage is chemically labile like that of an ester or thioester. The active-site thiol group in glyceraldehyde-3-phosphate dehydrogenase reacts with ***p*-chloromercuribenzoate** to form a mercuribenzoate adduct of the enzyme. Such adducts are not reversed by dialysis or by addition of substrate. Double-reciprocal plots show the characteristic pattern for noncompetitive inhibition.

Competitive inhibitors are inhibitors whose action can be reversed by increasing amount of substrate. Competitive inhibitors are structurally similar to the substrate and bind at the substrate-binding site, thus competing with the substrate for the enzyme. Once bound, the enzyme cannot convert the inhibitor to products. Increasing substrate concentrations will displace the reversibly bound inhibitor by the law of mass action. For example, in the **succinate dehydrogenase** reaction, malonate is structurally similar to succinate and is a competitive inhibitor.

Most modern drug therapy is based on the concepts of enzyme inhibition that were described in the previous section. Drugs are designed to inhibit a specific enzyme in a metabolic pathway. This application is most easily appreciated with antiviral, antibacterial, and antitumor drugs that are administered to the patient under conditions of limited toxicity. Such toxicity is often unavoidable because, with the exception of cell wall biosynthesis in bacteria, there are few critical metabolic pathways that are unique to tumors, viruses, or bacteria. Hence drugs that kill these organisms will often kill host cells. The one characteristic that can be taken advantage of is the comparatively short generation time of the undesirable organisms. They are much more sensitive to antimetabolites and in particular those that inhibit enzymes involved in replication. Antimetabolites are compounds with some structural difference from the natural substrate. In subsequent chapters, numerous examples of **antimetabolites** will be described.

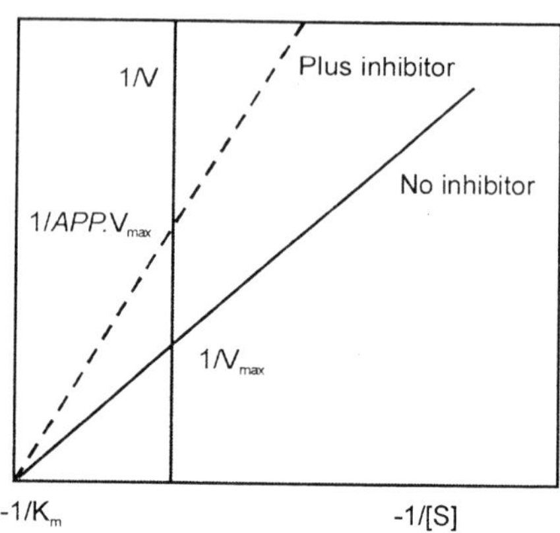

Double-reciprocal plot for an enzyme subject to reversible noncompetitive inhibition.

5.6 Enzyme Specifity

Enzymes are the most specific catalysts known, as regards the substrate and the type of reaction undergone by substrate. Specificity resides in the substrate binding site on the enzyme surface. The tertiary structure of the enzyme is folded in such a way as to create a region that has the correct molecular dimensions, topology, and the optimal alignment of counter ionic groups and hydrophobic regions to accommodate a specific substrate.

The tolerances in the active site are so small that usually only one isomer of a diastereomeric pair will bind. For example, D-amino acid oxidase will bind only D-amino acids but not L-amino acids. Some enzymes show absolute specificity for substrate. Others have broader specificity and will accept several different analogs of a specific substrate. For example, hexokinase catalyzes the phosphorylation of glucose, mannose, fructose, glucosamine, and 2-deoxyglucose, but at different rates. Glucokinase, on the other hand, is specific for glucose.

The specificity of the reaction rests in the active site and the amino acids that participate in the bond-making and bond-breaking phase of catalysis various models have been proposed to explain the substrate specificity of enzymes. The first proposal was the **"lock-and-key" model**, in which a negative impression of the substrate is considered to exist on the enzyme surface. Substrate fits in this binding site just as a key fits into the proper lock or a hand into the proper sized glove. Hydrogen and ionic bonding and hydrophobic interactions contribute in binding substrate to the binding site. This model gives a rigid picture of the enzyme and cannot account for the effect of allosteric ligands.

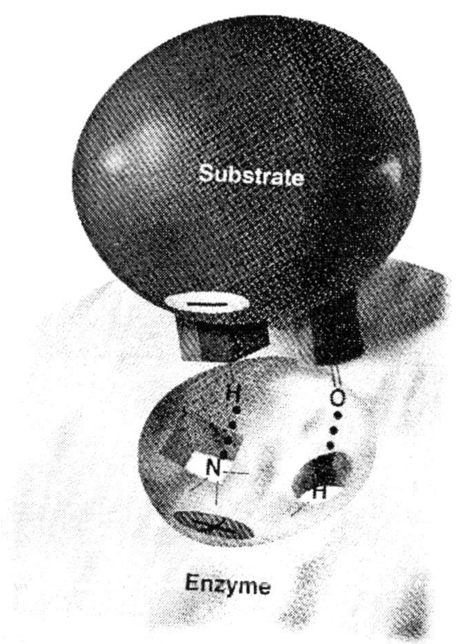

Lock-and-key model of the enzyme-binding site.

A more flexible model of the binding site is provided by the **induced fit model** in which the binding and active sites are not fully performed. The essential elements of the binding site are present to the extent that the correct substrate can position itself properly. Interaction of substrate with enzyme induces a conformational change in the enzyme, resulting in the formation of a stronger binding site and the repositioning of the appropriate amino acids to form the active site. There is excellent X-ray evidence for this model with carboxypeptidase A. The hexokinase essentially closes around the glucose to bring the active-site residues into proximity with the glucose.

Induced fit combined with **substrate strain** accounts for more experimental observations concerning enzyme action than other models. In this model substrate is "strained" toward product formation by an induced conformational transition of the enzyme.

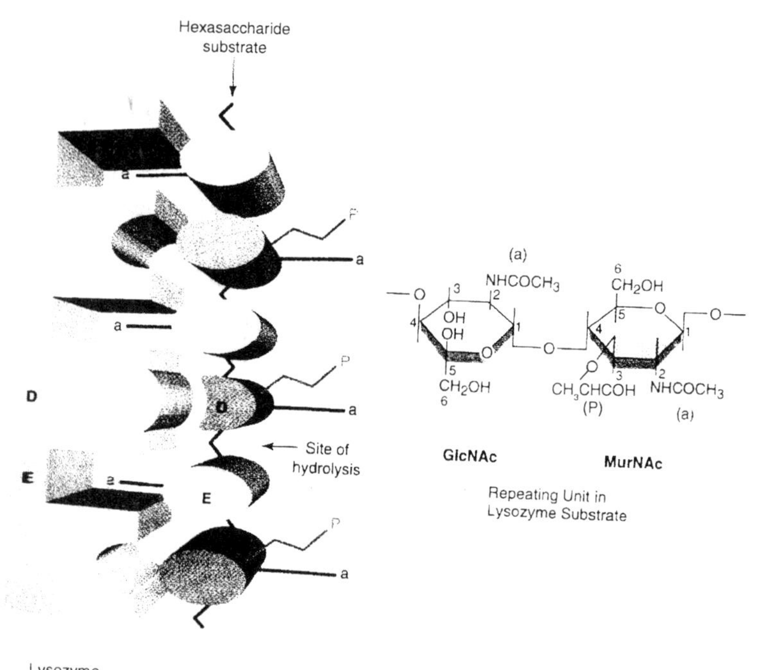

Hexasaccharide binding at active site of lysozyme.

A good example of enzyme induced substrate strain is that of **lysozyme** in which the conformation of the sugar residue "D" at which bond breaking occurs is strained from the stable chair to the unstable half-chair conformation upon binding. The concept of substrate strain explains the role of the enzyme in increasing the rate of reaction.

Not only are enzymes able to distinguish between isomers of the substrate, but they are able to distinguish between two equivalent atoms in a symmetrical molecule. For example, glycerol kinase distinguishes between configurations of H and OH on C-2 in the symmetric substrate glycerol, so that only the asymmetric product L-glycerol 3-phosphate is formed. These **prochiral substrates** have two identical substituents and two additional but dissimilar groups on the same carbon ($C_{aa'bd}$),

```
(a)  CH2OH                              CH2OH
      |                                  |
     CHOH (b,d) --glycerol kinase-->   HOCH        + ADP
      |              ATP                 |
(a') CH2OH                              CH2OPO3^2-

     Glycerol                    L-Glycerol 3-phosphate
```

Prochiral substrates possess no optical activity but can be converted to chiral compounds, that is, ones that possess an asymmetric center. The explanation for this enigma is provided if the enzyme binds the two dissimilar groups at specific sites and only one of the two similar substituents is able to bind at the active site. Thus the enzyme is able to recognize only one specific orientation of the symmetrical molecule. Asymmetry is produced in the product by modification of one side of the bound substrate. A minimum of three different binding sites on the enzyme surface is required to distinguish between identical groups on a prochiral substrate.

5.7 Mechanism of Catalysis

All chemical reactions have a potential energy barrier that must be overcome before reactants can be converted to products. In the gas phase the reactant molecules can be given enough kinetic energy by heating them so that collisions result in product formation. The same is true with solutions. However, a well-controlled body temperature of 37°C does not allow temperature to be increased to accelerate the reaction, and 37°C is not warm enough to provide the reaction rates required for fast-moving species of animals. Enzymes employ other means of overcoming the barrier to reaction.

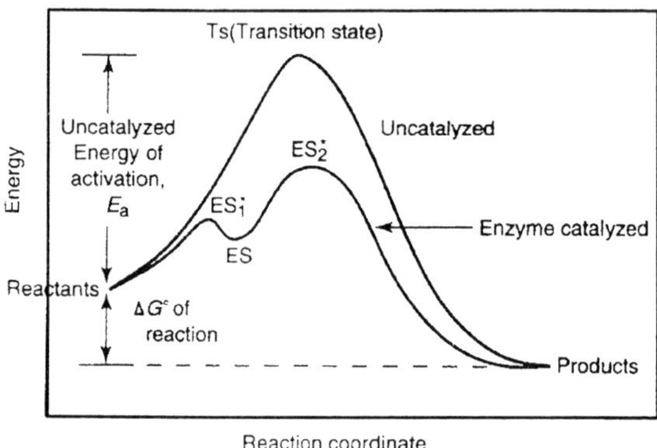

Energy diagrams for catalyzed versus noncatalyzed reactions.

Diagrams for catalyzed and noncatalyzed reactions are shown. The energy barrier represented by the uncatalyzed curve in figure is a measure of the **activation energy,** E_a, required for the reaction to occur. The reaction coordinate is simply the pathway in terms of bond stretching between reactants and products. At the apex of the energy barrier is the activated complex known as the **transition state,** Ts, that represents the reactants in their activated state. In this state reactants are in an intermediate stage along the reaction pathway and cannot be identified as starting material or products.

Now let us take the example of ethyle acetate.

$$CH_3-\overset{\overset{\large O}{\|}}{C}-O-CH_2-CH_3 \xrightarrow{H_2O} CH_3-CH_2-OH + CH_3-\overset{\overset{\large O}{\|}}{C}-OH$$

the Ts might look like

$$\left[CH_3-\underset{\underset{H\diagup\ \ \diagdown H}{\overset{\vdots}{O}}}{\overset{\overset{O^-}{|}}{C}}\cdots O-CH_2-CH_3 \right]$$

The transition state complex can break down to products or go back to reactants. The Ts is not an intermediate and cannot be isolated! In the case of the enzyme catalyzed reaction the energy of the reactants and products is no different than in the uncatalyzed reaction. Enzymes do not change the thermodynamics of the system but they do change pathway for reaching the final state.

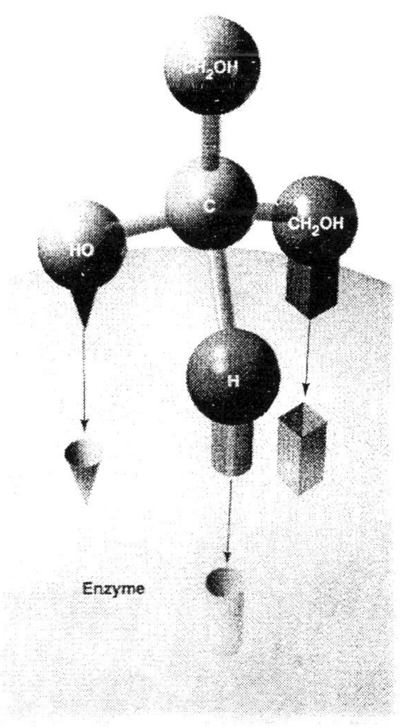

Three-point attachment of a symmetrical substrate to an asymmetric substrate-binding site.

Now let us try to learn about covalent catalysis.

In **covalent catalysis,** the attack of a nucleophilic (negatively charged) or electrophilic (positively charged) group in the enzyme active site upon the substrate results in covalent binding of the substrate to the enzyme as an intermediate in the reaction sequence. Enzyme-bound coenzymes often form covalent bonds with the substrate. For example, in the **transaminases,** the amino acid substrate forms a Schiff base with enzyme-bound pyridoxal phosphate. In all cases of covalent catalysis, the enzyme- or coenzyme bound substrate is more labile than the original substrate. The enzyme-substrate adduct represents one of the valleys on the energy profile.

Serine proteases, such as trypsin, chymotrypsin, and thrombin, are good representatives of the covalent catalytic mechanism. Acylated enzyme has been isolated in the case of chymotrypsin. Covalent catalysis is assisted by acid-base catalysis in these particular enzymes. In chymotrypsin the attacking nucleophile is Ser 195, which is not dissociated at pH 7.4 and a mechanism for ionizing this very basic group is required. It is now thought that in the anhydrous milieu of the active site, Ser 195 and His 57 have similar p*K* values and that the negative charge on Asp 102 stabilizes the transfer of the proton from the OH of Ser 195 to N3 of His 57. The resulting serine alkoxide attacks the carbonyl carbon of the peptide bond, releasing the amino-terminal end of the protein and forming an **acylated enzyme** intermediate (through Ser 195). The acylated enzyme is then cleaved by reversal of the reaction sequence, but with water as the nucleophile rather than Ser 195. Chemical evidence indicates the formation of two tetrahedral intermediates, one preceding the formation of the acylated enzyme and one following the attack of water on the acyl-enzyme.

The extent of disorder in a system can be defined, with the help of Entropy. Entropy is a thermodynamic term which is designated as ΔS.

At equilibrium, entropy is maximal. For example, in solution two reactants A and B exist in many different orientations. The chances of A and B coming together with the correct geometric orientation and with enough energy to react is small at 37°C and in dilute solution. However, if an enzyme with two high-affinity binding sites for A and B is introduced into the dilute solution of these reactants, A and B will be bound to the enzyme in the correct orientation for the reaction to occur. They will be bound with the correct stoichiometry, and the effective concentration of the reactants will be increased on the enzyme surface, all of which will contribute to an increased rate of reaction.

When correctly positioned and bound on the enzyme surface, the substrates may be "strained' toward the transition state.

Number of external Parameters including pH, temperature, and salt concentration affect enzyme activity. These effects are probably not important in vivo under normal conditions but are very important in setting up enzyme assays in vitro to measure enzyme activity in samples of a patients plasma or tissue.

5.8 Applications of Enzymes

Enzymes play a major role in modern **biotechnological industries** and are exploited in the laundry, food and drugs industries, the products of which make immense contributions to the quality of human life. There is nevertheless a constant demand for better, safer and more efficient products. Several important strategies are being developed to create improved enzymes. Immunoglobulin molecules, which are naturally devoid of catalytic powers, may be designed as **abzymes** to modify bound antigens. The use of gene cloning techniques permits the introduction of specific mutations into critical regions of the polypeptide chain and influences the functioning of the enzyme. This process, called **site-directed mutagenesis,** has been employed to modify the specificity of enzymic catalysis. Another advancing technique is the splicing of genes from different sources to code novel **hybrid enzymes.** Hybrid enzymes may be developed that create new specificities or enhance reaction rates.

The principles of enzymology are applied in the clinical laboratory in measurement of plasma or tissue enzyme activities and concentrations of substrates in patients. The rationale for measuring plasma enzyme activities is based on the premise that changes in activities reflect changes that have occurred in a specific tissue or organ. Plasma enzymes are of two types: (1) one type is present in the highest concentration, is specific to plasma, and has a functional role in plasma; and (2) the second is normally present at very low levels and plays no functional role in the plasma. The former includes the enzymes associated with blood coagulation (e.g., thrombin), fibrin dissolution (plasmin), and processing of chylomicrons (lipoprotein lipase).

In disease of tissues and organs, the nonplasma-specific enzymes are most important. Normally, the plasma levels of these enzymes are low to absent. A disease process may cause changes in cell membrane permeability or increased cell death, resulting in release of intracellular enzymes into the plasma. When permeability changes, those enzymes of lower molecular weight will appear in the plasma first and the greater the concentration gradient between intra- and extracellular levels, the more rapidly the enzyme diffuses out. Cytosolic enzymes will appear in the plasma before mitochondrial enzymes, and the greater the quantity of tissue damaged, the larger the increase in plasma level. The nonplasma-specific enzymes will be cleared from the plasma at varying rates, which depend on the stability of the enzyme and its uptake by the reticuloendothelial system.

In the diagnosis of specific organ involvement in a disease process it would be ideal if enzymes unique to each organ could

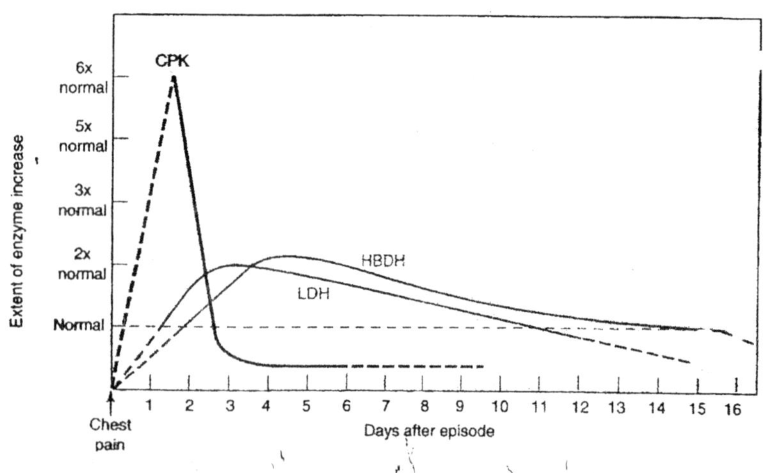

Kinetics of release of cardiac enzymes into serum following a myocardial infarction.

be identified. This is unlikely because the metabolic processes of various organs are very similar. Alcohol dehydrogenase of the liver and acid phosphatase of the prostate are useful for specific identification of disease in these organs. Other than these two examples, there are few enzymes that are tissue or organ specific. However, the ratio of various enzymes does vary from tissue to tissue. This fact, combined with a study of the kinetics of appearance and disappearance of particular enzymes in plasma, allows a diagnosis of specific organ involvement to be made.

Now let us learn about some other uses of enzymes. Clinical Analyzers use immobilized enzymes as reagents.

Enzymes are used as chemical reagents in desk-top clinical analyzers in offices or for screening purposes in shopping centers and malls. For example, screening tests for cholesterol and triacylglycerols can be completed in a few minutes using 10 μL of plasma. The active components in the assay system are cholesterol oxidase for the cholesterol determination and lipase for the triacylglycerols. The enzymes are immobilized in a bilayer along with the necessary buffer salts, cofactors or cosubstrates, and indicator reagents. The ingredients are arranged in a multilayered vehicle the size and thickness of a 35-mm slide. The plasma sample provides the substrate and water necessary to activate the system. In the case of cholesterol oxidase, hydrogen peroxide is a product that subsequently oxidizes a colorless dye to a colored product that is measured by reflectance spectroscopy. Peroxidase is included in the reagents to catalyze the latter exaction.

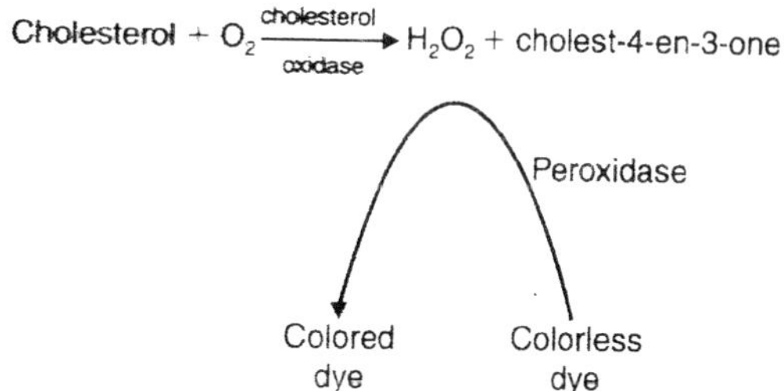

Each slide packet is constructed to measure a specific substance or enzyme and is stored in the cold for use as needed.

In many cases the slide packet contains several enzymes in a coupled assay system that eventually generates a reduced nucleotide or a colored dye that can be measured spectroscopically. This technology has been made possible, in part, by the fact that the enzymes involved are stabilized when bound to immobilized matrices and are stored in the dry state or in the presence of a stabilizing solvent such as glycerol.

Modern clinical chemistry has benefited from the marriage of enzyme chemistry and immunology. Antibodies specific to a protein antigen are coupled to an indicator enzyme such as horseradish peroxidase to generate a very specific and sensitive assay. After binding of the peroxidase-coupled antibody to the antigen, the peroxidase is used to generate a coloured product that is measurable and whose concentration is related to the amount of antigen in a sample. Because of the catalytic nature of the enzyme the system greatly amplifies the signal. This assay has been given the acronym ELISA for enzyme-linked immunoadsorbent assay.

Application of these principles is demonstrated by an assay for **human immunodeficiency virus (HIV)** coat protein antigens. This virus can lead to development of **acquired immunodeficiency syndrome (AIDS).** Antibodies are prepared in a rabbit against HIV coat proteins. In addition, a reporter antibody is prepared in a goat against rabbit IgG directed against the HIV protein. To this goat anti-rabbit IgG is linked the enzyme, horseradish peroxidase. The test for the virus is performed by incubating patient serum in a polystyrene dish that binds the proteins in the serum sample. Any free protein-binding sites remaining on the dish after incubation with patient serum are then covered by incubating with a nonspecific protein like bovine serum albumin. Next, the rabbit IgG antibody against the HIV protein is incubated in the dish during which time the IgG attaches to any HIV coat proteins that are attached to the polystyrene dish. All unbound rabbit IgG is washed out with buffer. The goat anti-rabbit IgG-peroxidase is now placed in the dish where it binds to any rabbit IgG attached to the dish via the HIV viral coat protein. Unattached antibody-peroxidase is washed out. Peroxidase substrates are added and the amount of colour developed in a given time period is a measurement of the amount of HIV coat protein present in a given volume of patient plasma when compared against a standard curve. This assay amplifies the signal because of the catalytic nature of the reporter group, the enzyme peroxidase. Such amplified enzyme assays allow the measurement of remarkably small amounts of antigents.

6

Lipid Metabolism

6.1 Introduction

Lipids have important functions in membrane structure and cellular metabolism. Lipids, excluding terpenes and steroids, contain fatty acid moeties. **Fatty acids** are therefore important as constituents of more complex molecules, e.g. phospho-acylglycerols, and are major sources of energy for most cells. In humans and mammals, small amounts of fatty acids are stored within most cells and larger amounts are generally found in specialized storage tissues, e.g. adipose tissue. The major storage form of fatty acids is **triacylglycerols,** delivered from the diet or synthesized intracellularly. Triacylglycerols are anhydrous and can release more energy per unit weight than hydrated storage carbo-hydrates. For this reason, during periods of excess food consumption, energy is stored as fat. Fatty acids supply about 40% of the calorific requirements of man on a normal diet. This figure may increase to almost 100% during periods of fasting and starvation. Fatty acids constitute the major energy supply during the migration of birds and the hybernation of animals.

As the human body builds and renews its structures, obtains and stores energy, and performs its various functions, there are many circumstances in which it is essential to use molecules or parts of molecules that do not associate with water. This property of being **nonpolar** and **hydrophobic** is the defining characteristic of substances classed as **lipids.** Most of these are molecules that contain or are derived from **fatty acids.** In the early stages of biochemical research, lipids were not investigated as intensively as other body constituents, largely because techniques for studying aqueous systems were easier to develop. This benign neglect led to assumptions that lipids were relatively inert and their metabolism was of lesser importance than that of carbohydrates, for instance.

As the methodology for analyzing lipid metabolism developed, it became evident that fatty acids and their derivatives had at least two major roles in the body. On the one hand, oxidation of fatty acids was shown to be a major means of metabolic energy production. It also became clear that their storage in the form of **triacylglycerols** was more efficient and quantitatively more important than storage of carbohydrates as glycogen. On the other hand, as details of the chemistry of biological structures were defined, hydrophobic structures were found to be largely composed of fatty acids and their derivatives. Thus the major separation of cells and subcellular structures into separate aqueous compartments is accomplished with membranes whose hydrophobic characteristics are largely supplied by the fatty acid moieties of complex lipids. These latter compounds contain constituents other than fatty acids and glycerol. They frequently have significant covalently bound hydrophilic moieties, notably carbohydrates in the glycolipids and organic phosphate esters in phospholipids.

Lipids have several other quantitatively less important roles, which are nonetheless of great functional significance. These include the use of surface active properties of some complex lipids for specific functions, such as maintenance of lung alveolar integrity and solubilization of nonpolar substances in body fluids. In addition, several classes of lipids, for example, steroid hormones and prostaglandins, have highly potent and specific physiological roles in control of metabolic processes. The metabolism of fatty acids and triacylglycerols is so crucial to proper functioning of the human body that imbalances and deficiencies in these processes can have serious pathological consequences. Disease states related to fatty acid and triacylglycerol metabolism include obesity, diabetes, ketoacidosis, and abnormalities in transport of lipids in blood. In addition, some unique deficiencies have been found, such as Refsum's disease and familial hypercholesterolemia, which have helped to elucidate some pathways in lipid metabolism.

In 1905, a key study on the **oxidation of fatty acids** was reported. **Knoop** conducted experiments in which dogs were fed **ω–phenyl derivatives** (a hydrogen atom of the terminal methyl group was substituted by a phenyl ring) of long-chain fatty acids with either an even or odd number of carbon atoms. The excretion products were the glycine conjugates of phenylacetate or benzoate called **phenylacetylglycine** (from even numbers) **and benzoylglycine** (from odd numbers). Knoop therefore concluded that the degradation of fatty acids must involve the successive removal of two-carbon units which he considered to be **acetate.**

The elucidation of the details of the process was delayed for 46 years until the discovery of coenzyme A (CoA). This discovery stimulated research and within the following 2 years, acetyl-CoA was identified as the major product of the *β*-oxidation pathway, the fundamental sequence of the reactions and the intracellular location of the pathway were established.

6.2 Fatty Acids and their Sources

Fatty acids consist of an **alkyl chain** with a terminal carboxyl group, the basic formula of completely saturated species being CH_3—$(CH_2)_n$—COOH. The important fatty acids for humans have relatively simple structures, although in some organisms they may be quite complex, containing cyclopropane rings or extensive branching. Unsaturation occurs commonly in human fatty acids, with up to six double bonds per chain, these being almost always of the cis configuration. If there is more than one double bond per molecule, they are always separated by a **methylene** (—CH_2—) **group.** The most common fatty acids in biological systems have an even number of carbon atoms, although some organisms do synthesize those with an odd number of carbon atoms. Humans can use the latter for energy but incorporate them into complex lipids to a minimal degree.

A few fatty acids with an α–OH group are produced and used structurally by humans. However, more oxidized forms are normally produced only as metabolic intermediates during energy production or for specific physiological activity in the case of prostaglandins and thromboxanes. Some animals, including humans, also produce relatively simple **branched-chain acids,** branching being limited to methyl groups along the chain at one or more positions. These are apparently produced to contribute specific physical properties to some secretions and structures. For instance, large amounts of branched-chain fatty acids, particularly isovaleric acid, occur in lipids of echo-locating structures in marine mammals. Elucidation of the role of these compounds in sound focusing should be fascinating.

Most fatty acids in humans have C16, C18, or C20 atoms, but there are several with longer chains that occur principally in lipids of the nervous system. These include nervonic acid and a C22 acid with six double bonds.

The most abundant fatty acids have common names that have been accepted for use in the official nomenclature. Example are given in. The approved abbreviations consist of the number of carbon atoms followed, after a colon, by the number of double bonds. Carbon atoms are numbered with the carboxyl carbon as number 1, and double bond locations are designated by the number of the carbon atom on the carboxyl side of it. These designations of double bonds are in parentheses after the rest of the symbol.

Fatty acids occur primarily as esters of glycerol, when they are stored for future use. Compounds with one (**monoacylglycerols**) or two (**diacylglycerols**) acids esterified are present only in relatively minor amounts and occur largely as metabolic intermediates in biosynthesis and degradation of glycerol-containing lipids. Most fatty acids in humans exist as **triacylglycerols,** in which all three hydroxyl groups on glycerol are esterified with a fatty acid. These compounds have been called "neutral fats or triglycerides." There are other types of "neutral fats" in the body, and the terms "triglyceride" "monoglyceride," and "diglyceride" are chemically incorrect and should not be used.

One significant property of fatty acids and triacylglycerols is their lack of affinity for water. Long hydrocarbon chains have negligible possibility for hydrogen bonding. Acids, whether unesterified or in a complex lipid, have a much greater tendency to associate with each other or other hydrophobic structures, such as sterols and hydrophobic side chains of amino acids, than they do with water or polar organic compounds. This hydrophobic character is essential for construction of complex biological structures such as membranes.

The **hydrophobic nature** of triacylglycerols and their highly reduced state make them efficient compounds in comparison to glycogen for storing energy. Three points deserve emphasis. First, on a weight basis pure triacylglycerols yield near two and one-half times the amount of ATP on complete oxidation than does pure glycogen. Second, triacylglycerols can be stored without associated water, whereas glycogen is very hydrophilic and binds about twice its weight of water when stored in tissues. Thus the equivalent amount of metabolically recoverable energy stored as hydrated glycogen would weigh about four times as much as if it were stored as triacylglycerols. Third, the average 70-kg person stores about 350 g of carbohydrate as liver and muscle glycogen. This represents about 1400 kcal of available energy, barely enough to sustain bodily functions for 24 hours of fasting. By contrast, a normal complement of fat stores will provide sufficient energy to allow several weeks of survival during told food deprivation.

In humans most of the fatty acids are either saturated or contain only one double bond. Although they are readily catabolized by appropriate enzymes and cofactors, they are fairly inert chemically. The highly unsaturated fatty acids in tissues are much more susceptible to oxidation.

The fatty acids diffuse through the plasma membranes and are transported in the blood circulation as a **serum albumin-acylate complex** to the tissues requiring fatty acids for oxidation. At these tissues, the fatty acids dissociate from the complex and enter the tissue cells by diffusion. This aspect of fatty acid transport is governed by the ratio of the concentration of albumin-bound fatty acid to the intracellular concentration of fatty acid. The long-chain fatty acids delivered to the tissue cells undergo cytosolic activation by an **acyl-CoA synthase** (also called **fatty acid-CoA ligase**) before entry into the mitochondrion for oxidation.

Fatty acids are very much needed by the human body for energy and for construction of hydrophobic parts of biomolecules.

Now let us study about the sources of fatty acids.

Various animals and vegetable lipids are ingested, hydrolyzed at least partially by digestive enzymes, and absorbed through the intestinal mucosa to be distributed through the body, first in the lymphatic system and then in the bloodstream. To a large extent dietary supply governs the composition of fatty acids in body lipids. Metabolic processes in various tissues modify both dietary and *de novo* synthesized fatty acids to produce nearly all the required structures. With one exception, the actual composition of fatty acids supplied in the diet is relatively unimportant. This exception involves the need for appropriate proportions of relatively highly unsaturated fatty acids because many higher mammals, including humans, are unable to synthesize fatty acids with double bonds near the methyl end of the molecule. Certain **polyunsaturated acids** with double bonds within the last seven linkages toward the methyl end are essential for specific functions.

Humans can synthesize all of the fatty acids they need from palmitate except the essential, polyunsaturated fatty acids. These syntheses involve a variety of enzyme systems in a number of locations. Palmitate produced by fatty acid synthase is modified by three processes: elongation, desaturation, and hydroxylation.

In mammals **elongation of fatty acids** occurs in either the endoplasmic reticulum or mitochondria; the processes are slightly different in these two loci. In the endoplasmic reticulum the sequence of reactions is similar to that occurring in the cytosolic

The Regulation of Fatty Acid Synthesis can be studied as follows.

Enzyme		*Regulatory Agent*	*Effect*
		Palmitate Biosynthesis	
Acetyl-CoA carboxylase	Short term	Citrate	Allosteric activation
		C16-C18 acyl CoAs	Allosteric inhibition
		Insulin	Stimulation
		Glucagon	Inhibition
		cAMP-mediated phosphorylation	Inhibition
		Dephosphorylation	Stimulation
	Long term	High-carbohydrate diet	Stimulation by increased enzyme synthesis
		Fat-free diet	Stimulation by increased enzyme synthesis
		High-fat diet	Inhibition by decreased enzyme synthesis
		Fasting	Inhibition by decreased enzyme synthesis
		Glucagon	Inhibition by decreased enzyme synthesis
Fatty acid synthase		Phosphorylated sugars	Allosteric activation
		High-carbohydrate diet	Stimulation by increased enzyme synthesis
		Fat-free diet	Stimulation by increased enzyme synthesis
		High-fat diet	Inhibition by decreased enzyme synthesis
		Fasting	Inhibition by decreased enzyme synthesis
		Glucagon	Inhibition by decreased enzyme synthesis
		Biosynthesis of Fatty Acids Other than Palmitate	
Fatty acid synthase		High ratio of methylmalonyl CoA/malonyl CoA	Increased synthesis of methylated fatty acids
		Thioesterase cofactor	Termination of synthesis with short-chain product
Stearyl CoA desaturase		Various hormones	Stimulation of unsaturated fatty acid synthesis by increased enzyme synthesis
		Dietary polyunsaturated fatty acids.	Decreased activity

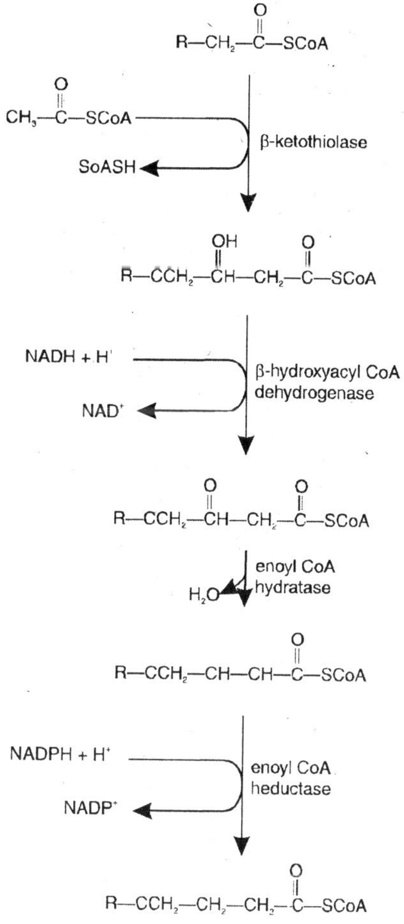

R—CH$_2$—C(=O)—SCoA

CH_3—C(=O)—SCoA → SoASH (β-ketothiolase)

R—CCH$_2$—CH(OH)—CH$_2$—C(=O)—SCoA

NADH + H$^+$ → NAD$^+$ (β-hydroxyacyl CoA dehydrogenase)

R—CCH$_2$—CH(O)—CH$_2$—C(=O)—SCoA

→ H_2O (enoyl CoA hydratase)

R—CCH$_2$—CH—CH—C(=O)—SCoA

NADPH + H$^+$ → NADP$^+$ (enoyl CoA heductase)

R—CCH$_2$—CH$_2$—CH$_2$—C(=O)—SCoA

Mitochondrial elongation of fatty acids.

fatty acid synthase with malonyl CoA as the source of two-carbon units and NADPH providing the reducing power. The preferred substrate for elongation is palmitoyl CoA. In contrast to palmitate synthesis, intermediates in subsequent reactions are CoA esters rather than attached to a protein, suggesting that the process is carried out by separate enzymes rather than a complex like fatty acid synthase. In most tissues this elongation system in the endoplasmic reticulum converts palmitate to stearate almost exclusively. Brain, however, contains one or more additional elongation systems, which synthesize longer chain acids (up to C24) needed for brain lipids. These other systems also use malonyl CoA as substrate.

The elongation system in mitochondria differs in that acetyl CoA is the source of the added two-carbon units and both NADH and NADPH serve as reducing agents. This system operates by reversal of the pathway of fatty acid β–oxidation with the exception that NADPH-linked enoyl-CoA reductase (last step of elongation) replaces FAD-linked acyl-CoA dehydrogenase (first step in β-oxidation). The process has little activity with acyl CoA substrates of C16 atoms or longer, suggesting that it serves primarily in elongation of shorter chain species.

In higher animals **desaturation of fatty acids** occurs in the endoplasmic reticulum, and the oxidizing system used to introduce cis double bonds is significantly different from the main fatty acid oxidation process in mitochondria. The systems in endoplasmic reticulum have sometimes been termed **"mixed function oxidases"** because the enzymes simultaneously oxidize two substrates. In fatty acid desaturation one of these substrates is NADPH and the other is the fatty acid. Electrons from NADPH are transferred through a specific flavoprotein reductase and a cytochrome to "active" oxygen so that it will then oxidize the fatty acid. Although the complete mechanism has not been determined, this latter step may involve a hydroxylation. The three components of the system are the **desaturase enzyme, cytochrome b_5,** and **NADPH-cytochrome b_5 reductase.** The overall reaction is

$$\begin{aligned} & R{-}CH_2{-}CH_2{-}(CH_2)_7{-}COOH + NADPH + H^+ + O_2 \rightarrow \\ & \qquad R{-}CH{=}CH{-}(CH_2)_7{-}COOH + NADP^+ + 2\,H_2O \end{aligned}$$

The enzyme specificity is such that the R group must contain at least six carbon atoms.

The regulatory mechanisms that govern the conversion of palmitate to unsaturated fatty acids are largely unexplored. An important consideration is the control of the proportions of unsaturated fatty acids for proper maintenance of the physical state of stored triacylglycerols and membrane phospholipids. A committed step in the formation of unsaturated fatty acids from palmitate or stearate is introduction of the first double bond between C-9 and C-10 atoms by **stearoyl CoA desaturase** to produce palmitoleic or oleic acid, respectively. The activity of this enzyme and its synthesis are controlled by both dietary and hormonal mechanisms.

6.3 The Need of Fatty Acids for Energy Production

The degree of utilization of fatty acids for energy production varies considerably from tissue to tissue and depends to a significant extent on the metabolic status of the body, whether it

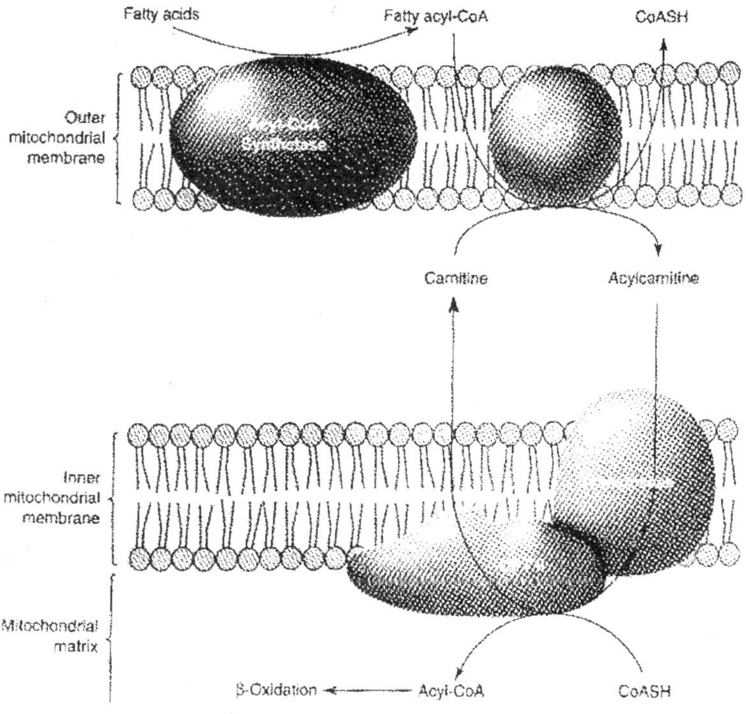

Mechanism for transfer of fatty acids from the cytosol through the inner mitochondrial membrane for oxidation.

is fed or fasted, exercising or at rest, and so on. For instance, nervous tissue oxidizes fatty acids to a minimal degree, if at all, but cardiac and skeletal muscle depend heavily on fatty acids as a major energy source. During prolonged fasting most tissues can use fatty acids or ketone bodies for their energy requirements.

For the most part, fatty acids are oxidized by a mechanism that is similar to, but not identical with, a reversal of the process of palmitate synthesis. That is, two-carbon fragments are removed sequentially from the carboxyl end of the acid after steps of **dehydrogenation, hydration,** and **oxidation** to form a *β*–keto acid, which is split by **thiolysis.** These processes take place while the acid is activated in a thioester linkage to the 4′-phosphopantetheine of CoA.

Now let us study in brief about β-oxidation pathway.

In β-oxidation pathway, a long chain fatty acid in the ionized form is activated in the cytosol by transmembranous acyl-CoA synthase, the coenzyme A-binding site of which is located on the cytosolic surface of the outer mitochondrial membrane.

Each set of oxidations resulting in production of a two-carbon fragment yields, in addition to acetyl CoA, one reduced flavoprotein and one NADH. In the oxidation of palmitoyl CoA, seven such cleavages take place, and in the last cleavage two acetyl CoA molecules are formed. Thus the products of *β*–oxidation of palmitate are eight acetyl CoAs, seven reduced flavoproteins, and seven NADH. Each of the reduced flavoproteins can yield two ATP and each NADH can yield three when oxidized by the electron transport chain, so the reduced nucleotides yield 35 ATP per palmitoyl CoA. Oxidation of each acetyl CoA through the TCA cycle yields 12 ATP, so the eight two-carbon fragments from a palmitate molecule produce 96 ATP. However, 2 ATP equivalents (1 ATP going to 1 AMP) were used to activate palmitate to palmitoyl CoA. Therefore each palmitic acid entering the cell from the action of lipoprotein lipase or from its combination with serum albumin can yield 129 ATP mol^{-1} by complete oxidation.

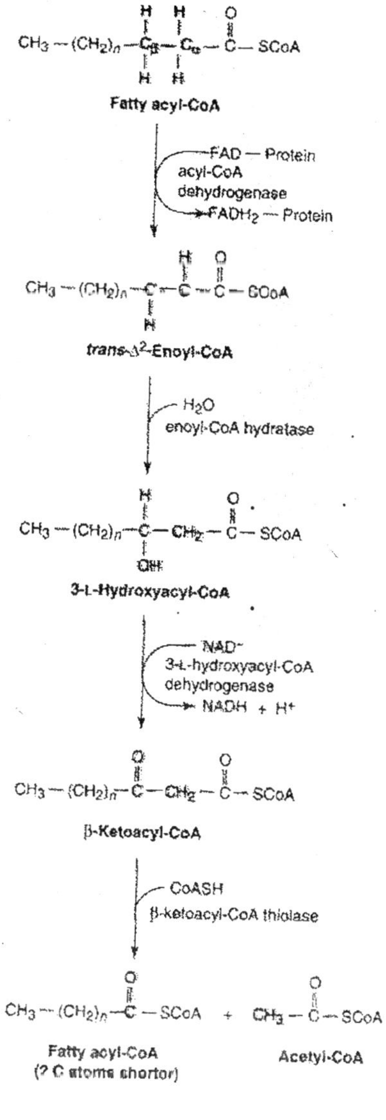

$CH_3-(CH_2)_n-C_\beta H_2-C_\alpha H_2-\overset{O}{\overset{\|}{C}}-SCoA$

Fatty acyl-CoA

↓ FAD — Protein → $FADH_2$ — Protein (acyl-CoA dehydrogenase)

$CH_3-(CH_2)_n-CH=CH-\overset{O}{\overset{\|}{C}}-SCoA$

***trans*-Δ^2-Enoyl-CoA**

↓ H_2O (enoyl-CoA hydratase)

$CH_3-(CH_2)_n-CH(OH)-CH_2-\overset{O}{\overset{\|}{C}}-SCoA$

3-L-Hydroxyacyl-CoA

↓ NAD^+ → $NADH + H^+$ (3-L-hydroxyacyl-CoA dehydrogenase)

$CH_3-(CH_2)_n-\overset{O}{\overset{\|}{C}}-CH_2-\overset{O}{\overset{\|}{C}}-SCoA$

β-Ketoacyl-CoA

↓ CoASH (β-ketoacyl-CoA thiolase)

$CH_3-(CH_2)_n-\overset{O}{\overset{\|}{C}}-SCoA \quad + \quad CH_3-\overset{O}{\overset{\|}{C}}-SCoA$

Fatty acyl-CoA (2 C atoms shorter) **Acetyl-CoA**

Pathway of fatty acid β–oxidation.

In living metabolic systems the reactions in a catabolic pathway are sometimes quite similar to those in a reversal of the corresponding anabolic sequence, but there are usually mechanisms that provide for separate control of the two schemes. This is true in the case of fatty acid biosynthesis and β-oxidation. They include separation by subcellular compartmentation (β-oxidation occurs in the mitochondria and palmitate biosynthesis in the cytosol) and use of different cofactors (NADPH in biosynthesis, FAD and NAD^+ in oxidation).

Comparison of Schemes for Biosynthesis and β–Oxidation of Palmitate

Parameter	*Biosynthesis*	*β-Oxidation*
Subcellular localization	Primarily cytosolic	Primarily mitochondrial
Phosphopantetheine-containing active carrier	Acyl carrier protein	Coenzyme A
Nature of small carbon fragment added or removed	C-1 and C-2 atoms of malonyl CoA after initial priming	Acetyl CoA
Nature of oxidation-reduction coenzyme	NADPH	FAD when saturated chain dehydrogenated, NAD^+ when hydroxy acid dehydrogenated
Stereochemical configuration of β-hydroxy intermediates	D-β-Hydroxy	L-β-Hydroxy
Energy equivalents yielded or utilized in interconversion of palmitate and acetyl CoA	7 ATP + 14 NADPH = 49 ATP equiv	7 $FADH_2$ + 7 NADH – 2 ATP = 33 ATP equiv.

6.4 Phospholipids

Lipid is a general term that describes substances that are relatively water insoluble and extractable by nonpolar solvents. Complex lipids of humans fall into one of two broad categories: nonpolar lipids, such as triacylglycerols and cholesterol esters, and polar lipids, which are amphipathic in that they contain both a hydrophobic domain and a hydrophilic region in the same molecule. This chapter discusses polar lipids including *phospholipids, sphingolipids,* and *eicosanoids.* The hydrophobic and hydrophilic regions are bridged by a glycerol moiety in glycerophospholipids and by sphingosine in sphingomyelin and glycosphingolipids. Triacylglycerol is confined largely to storage sites in adipose tissue, whereas polar lipids occur primarily in cellular membranes. Oils of soybean, oil palm, rapeseed, sunflower, cottonseed, and peanut account for 80% of worldwide plant oil production and consist mainly of palmitic, stearic, oleic, linoleic, and α-linolenic acids. Membranes generally contain 40% of their dry weight as lipid and 60% as protein.

Cell-cell recognition, phagocytosis, contact inhibition, and rejection of transplanted tissues and organs are all phenomena of medical significance that involve highly specific recognition sites on the surface of plasma membranes. Synthesis of the complex glycosphingolipids that play a role in these important biological events will be described. Glycolipids are worthy of study because ABO antigenic determinants of blood groups are primarily glycolipid in nature. In addition, various sphingolipids are the storage substances that accumulate in liver, spleen, kidney, or nervous tissue of persons suffering from certain genetic disorders called sphingolipidoses.

Phospholipids are polar, ionic lipids composed of 1,2-diacylglycerol and a phosphodiester bridge that links the glycerol backbone to some base, usually a nitrogenous one, such as choline, serine, or ethanolamine. The most abundant phospholipids in human tissues are **phosphatidylcholine** (also called lecithin), **phosphatidylethanolamine,** and **phosphatidylserine**. At physiologic pH, phosphatidylcholine and phosphatidylethanolamine have no net charge and exist as dipolar zwitterions, whereas phosphatidylserine has a net charge of –1, causing it to be an acidic phospholipid. Phosphatidylethanolamine (PE) is related to phosphatidylcholine in that trimethylation of PE produces lecithin. Most phospholipids contain more than one kind of fatty acid per molecule, so that a given class of phospholipids from any tissue actually represents a family of molecular species. Phosphatidylcholine (PC) contains mostly palmitic acid (16:0) or stearic acid (18:0) in the *sn*-1 position and primarily unsaturated C18 fatty acids oleic, linoleic, or α-linolenic in the *sn*-2 position. Phosphatidylethanolamine has the same saturated fatty acids as PC at the *sn*-1 position but contains more of the long-chain polyunsaturated fatty acids—namely, 18:2, 20:4, and 22:6—at the *sn*-2 position.

Phosphatidylinositol is an acidic phospholipid that occurs in mammalian membranes. Phosphatidylinositol is rather unusual because if often contains almost exclusively stearic acid in the *sn*-1 position and arachidonic acid (20:4) in the *sn*-2 position.

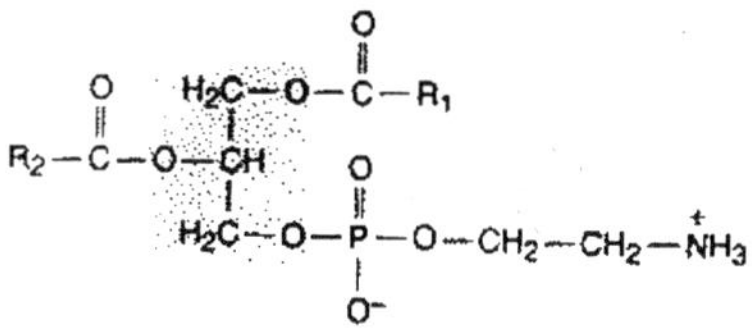

$$\begin{array}{l} H_2C-O-C(=O)-R_1 \\ R_2-C(=O)-O-CH \\ H_2C-O-P(=O)(O^-)-O-CH_2-CH_2-\overset{+}{N}H_3 \end{array}$$

Phosphatidylethanolamine

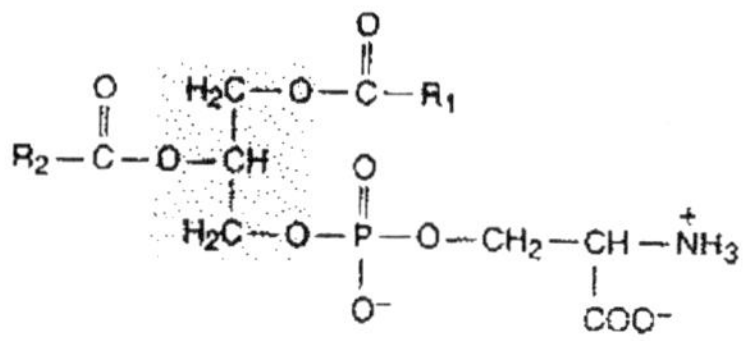

$$\begin{array}{l} H_2C-O-C(=O)-R_1 \\ R_2-C(=O)-O-CH \\ H_2C-O-P(=O)(O^-)-O-CH_2-CH(COO^-)-\overset{+}{N}H_3 \end{array}$$

Phosphatidylserine

$$\begin{array}{l} H_2C-O-C(=O)-R_1 \\ R_2-C(=O)-O-CH \\ H_2C-O-P(=O)(O^-)-O-CH_2-CH_2-\overset{+}{N}-(CH_3)_3 \end{array}$$

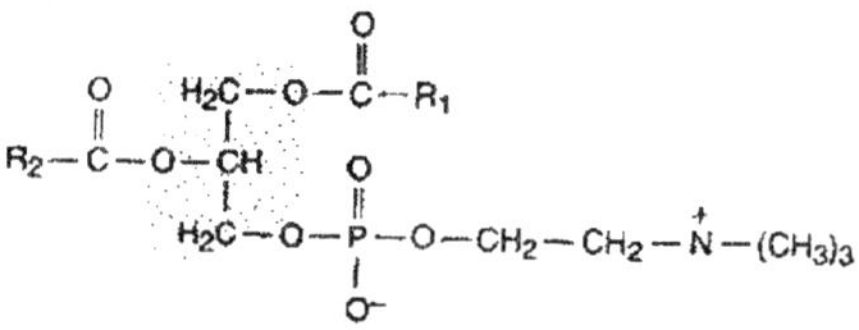

Phosphatidylcholine (lecithin)

Structures of some common phospholipids.

Another phospholipid comprised of a polyol polar head group is **phosphatidylglycerol** which occurs in relatively large amounts in mitochondrial membranes and pulmonary surfactant and is a precursor of cardiolipin mitochondrial membranes and pulmonary surfactant and is a precursor of **cardiolipin.** Phosphatidylglycerol and phosphatidylinositol both carry a formal charge of –1 at neutral pH and are therefore acidic lipids. Cardiolipin, a very acidic (charge, –2) phospholipid, is composed of two molecules of phosphatidic acid linked together covalently through a molecule of glycerol. It occurs primarily in the inner membrane of mitochondria and in bacterial membranes.

Phospholipids mentioned so far contain only *O*-acyl residues attached to glycerol. *O*-(1-Alkenyl) substituents occur at C-1 of the *sn*-glycerol in phosphoglycerides in combination with an *O*-acyl residue esterified to the C-2 position: compounds in this class are known as **plasmalogens**. Relatively large amounts of ethanolamine plasmalogen (also called plasmenylethanolamine) occur in myelin with lesser amounts in heart muscle where choline plasmalogen is abundant.

An unusual phospholipid called **"platelet-activating factor"** (PAF) is a major mediator of hypersensitivity, acute inflammatory reactions and anaphylactic shock. In hypersensitive individuals, cells of the polymorphonuclear (PMN) leukocyte family (basophils, neutrophils, and eosinophils), macrophages, and monocytes are coated with IgE molecules that are specific for a particular antigen (e.g., ragweed pollen and bee venom).

6.5 Phosphatidic Acid Biosynthesis

1-α-Phosphatidic acid (commonly called phosphatidic acid) and 1,2-diacyl-*sn*-glycerol are common intermediates in the pathways of phospholipid and triacylglycerol biosynthesis and both pathways share some of the same enzymes. Essentially all cells are capable of synthesizing phospholipids to some degree (except mature erythrocytes), whereas triacylglycerol biosynthesis occurs only in liver, adipose tissue, and intestine. In most tissues, the pathway for phosphatidic acid synthesis begins with α-glycerophosphate (*sn*-glycerol 3-phosphate). The most general source of

α-glycerophosphate, particularly in adipose tissue, is from reduction of the glycolytic intermediate, dihydroxyacetone phosphate, in the reaction catalyzed by α-glycerophosphate dehydrogenase:

$$\text{Dihydroxyacetone phosphate} + \text{NADH} + \text{H}^+ \rightleftharpoons \alpha\text{-glycerol 3-phosphate} + \text{NAD}^+$$

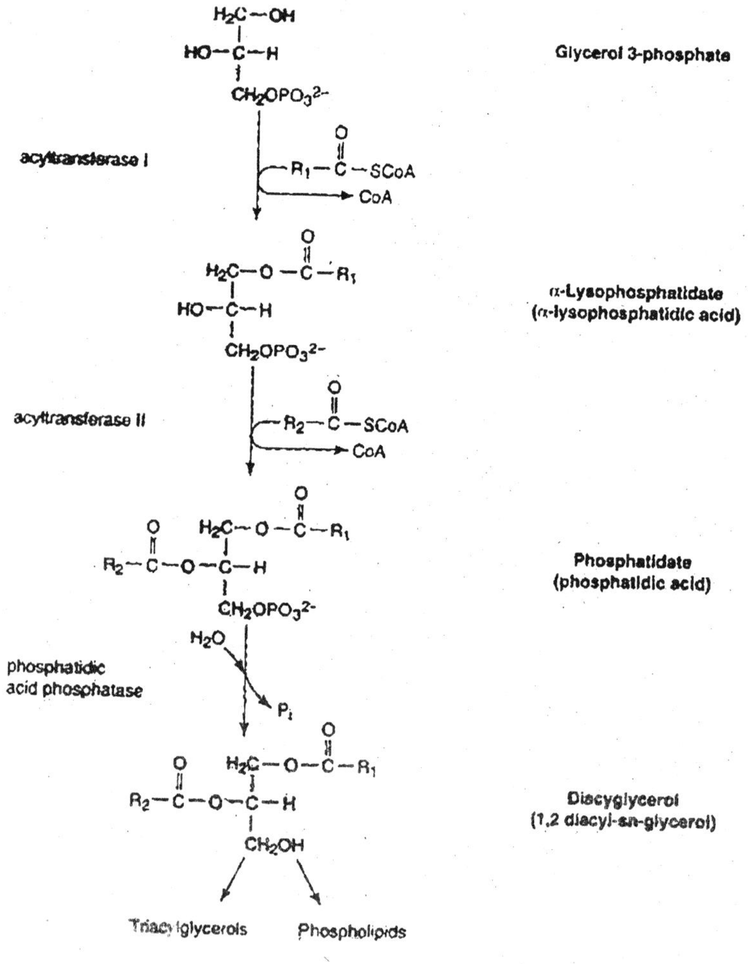

Glycerol 3-phosphate

acyltransferase I

α-Lysophosphatidate
(α-lysophosphatidic acid)

acyltransferase II

Phosphatidate
(phosphatidic acid)

phosphatidic
acid phosphatase

Diacyglycerol
(1,2 diacyl-sn-glycerol)

Triacylglycerols   Phospholipids

Phosphatidic acid biosynthesis from glycerol 3-phosphate and the role of phosphatidic acid phosphatase in synthesis of phospholipids and triacylglycerols.

A few specialized tissues, including liver and kidney, derive α-glycerophosphate by means of the glycerol kinase reaction:

$$\text{Glycerol} + \text{ATP} \xrightleftharpoons{Mg^{2+}} \alpha\text{-glycerol 3-phosphate} + \text{ADP}$$

The next two steps in phosphatidic acid biosynthesis involve stepwise transfer of long-chain fatty acyl groups from fatty acyl CoA. The first acyltransferase (I) is called **glycerol phosphate: acyltransferase** and attaches predominantly saturated fatty acids or oleic acid to the *sn*-1 to produce 1-acylglycerol phosphate or α-lysophosphatidic acid. The second enzyme (II), **1-acylglycerol phosphate:acyltransferase,** acylates the *sn*-2 position, usually with an unsaturated fatty acid. In both cases the donor of acyl groups is the CoA thioester derivative of the appropriate long-chain fatty acid.

The specificity of the two acyltransferases does not always match the fatty acid asymmetry found in the phospholipids of a particular cell. Remodeling reactions, discussed below, modify the fatty acid composition at C-1 and C-2 of the glycerol phosphate backbone.

Cytosolic phosphatidic acid phosphatase (also called phosphatidic acid phosphohydrolase) hydrolyzes phosphatidic acid (1,2-diacylglycerophosphate) that is generated on the endoplasmic reticulum, thereby yielding 1,2-diacyl-*sn*-glycerol that serves as the branch point in triacylglycerol and phospholipid synthesis. Phosphatidic acid can also be formed by a second pathway that begins with DHAP. This is usually an alternative supportive route used by some tissues to produce phosphatidic acid.

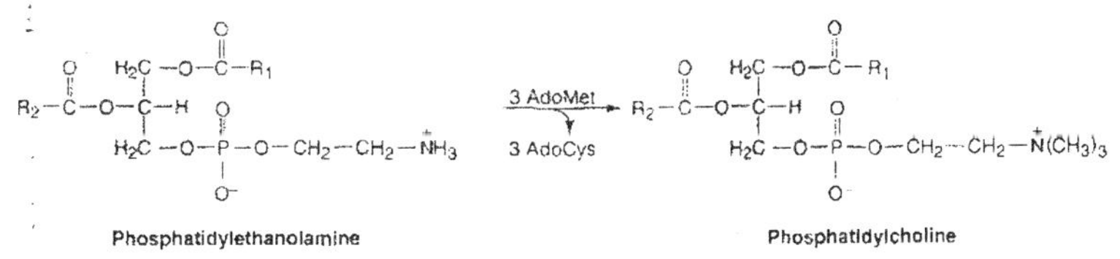

Phosphatidylethanolamine $\xrightarrow[\text{3 AdoCys}]{\text{3 AdoMet}}$ Phosphatidylcholine

Biosynthesis of phosphatidylcholine from phosphatidylethanolamine and *S*-adenosylmethionine (AdoMet) and *S*-adenosylhomocysteine (AdoCys).

6.6 Cholesterol

Cholesterol is an alicyclic compound whose structure includes: (1) the perhydrocyclopentanophenanthrene nucleus with its four fused rings: (2) a single hydroxyl group at C-3, (3) an unsaturated center between C-5 and C-6 atoms; (4) an eight-membered branched hydrocarbon chain attached to the D ring at position 17; and (5) a methyl group (designated C-19) attached at position 10 and another methyl group (designated C-18) attached at position 13.

In terms of physical properties, cholesterol is a lipid with very low solubility in water; at 25°C, the limit of solubility is approximately 0.2 mg/100 mL. The actual concentration of cholesterol in plasma of healthy people is usually 150-200 mg/100 mL; this value is almost twice the normal concentration of blood glucose. This high solubility of cholesterol in blood is due to plasma lipoproteins (mainly LDL and VLDL) that have the ability to bind and thereby solubilize large amounts of cholesterol. Actually, only about 30% of the total plasma cholesterol occurs free; approximately 70% of the cholesterol in plasma lipoproteins exists in the form of **cholesterol esters** where some long-chain fatty acid, usually linoleic acid, is attached by an ester bond to the OH group on C-3 of the A ring. The long-chain fatty acid residue enhances the hydrophobicity of cholesterol. Cholesterol is a ubiquitous and essential component of mammalian cell membranes.

Cholesterol is also abundant in bile where the normal concentration is 390 mg/100 mL. Only 4% of cholesterol in bile is esterified to a long-chain fatty acid. Bile does not contain appreciable amounts of lipoproteins and solubilization of free cholesterol is achieved in part by the detergent property of phospholipids present in bile that are produced in liver. A chronic disturbance in phospholipid metabolism in liver can result in deposition of cholesterol-rich gallstones. Bile salts, which are derivatives of cholesterol, also aid in keeping cholesterol in solution in bile. Cholesterol also appears to protect membranes of the gallbladder from potentially irritating or harmful effects of bile salts.

In the clinical laboratory, total cholesterol is estimated by the Liebermann-Burchard reaction. The proportions of free and esterified cholesterol can be determined by gas–liquid chromatography or reverse-phase high-pressure liquid chromatography (HPLC).

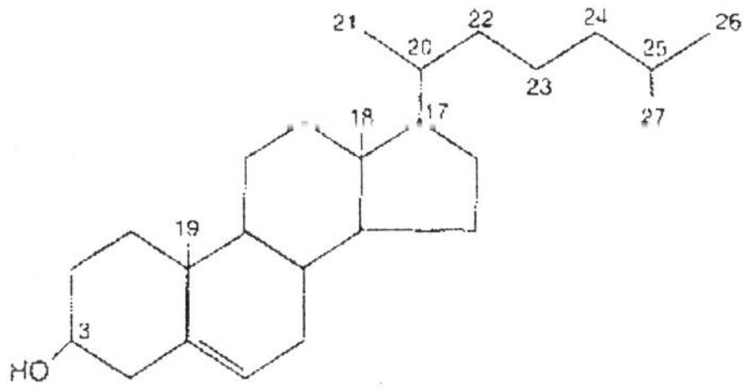

Structure of cholesterol (5-cholesten-3β-ol).

Cholesterol, derived from the diet or synthesized *de novo* in virtually all cells of humans, has a number of important roles. It is the major sterol in humans and a component of virtually all plasma and intracellular membranes. Cholesterol is especially abundant in myelinated structures of brain and central nervous system but is present in small amounts in the inner membrane of the mitochondrion. In contrast to the situation in plasma, most cholesterol in cellular membranes occurs in the free, unesterified form.

The first compound unique to cholesterol biosynthesis is mevalonic acid derived from acetyl CoA. Acetyl CoA can be obtained from several sources: (1) the β–oxidation of fatty acids; (2) the oxidation of ketogenic amino acids such as leucine and isoleucine; and (3) the pyruvate dehydrogenase reaction. Free acetate can be activated to its thioester derivative at the expense of ATP by **acetokinase,** also referred to as **acetate thiokinase:**

$$ATP + CH_3COO^- + CoASH \rightarrow CH_3 - \overset{\overset{\displaystyle O}{\|}}{C} - SCoA + AMP + PP_i$$

The first two reactions in cholesterol biosynthesis are shared by the pathway that produces ketone bodies. Two molecules of acetyl CoA condense to form acetoacetyl CoA in a reaction catalyzed by acetoacetyl CoA thiolase (acetyl CoA:acetyl CoA acetyltransferase):

$$CH_3-\overset{\overset{\displaystyle O}{\|}}{C}-SCoA + CH_3\overset{\overset{\displaystyle O}{\|}}{C}-SCoA \rightarrow CH_3-\overset{\overset{\displaystyle O}{\|}}{C}-CH_2-\overset{\overset{\displaystyle O}{\|}}{C}-SCoA + CoA$$

Formation of the carbon–carbon bond in acetoacetyl CoA in this reaction is favoured energetically by cleavage of a thioester bond and generation of free coenzyme A.

The next step introduces a third molecule of acetyl CoA into the cholesterol pathway and forms the branched-chain compound **3-hydroxy-3-methylglutaryl CoA** (HMG CoA). This condensation reaction is catalyzed by **HMG CoA synthase** (3-hydroxy-3-methylglutaryl CoA:acetoacetyl CoA lyase). Liver parenchymal cells contain two isoenzyme forms of HMG CoA synthase; one in the cytosol is involved in cholesterol synthesis, while the other has a mitochondrial location and functions in synthesis of ketone bodies. In the HMG CoA synthase reaction, an aldol condensation occurs between the methyl carbon of acetyl CoA and the β-carbonyl group of acetoacetyl CoA with the simultaneous hydrolysis of the thioester bond of acetyl CoA. The thioester bond in the original acetoacetyl CoA substrate molecule remains intact. HMG CoA can also be formed from oxidative degradation of the branched-chain amino acid leucine, through the intermediates 3-methylcrotonyl CoA and 3-methylglutaconyl CoA.

The step that produces the unique compound mevalonic acid from HMG CoA is catalyzed by the important microsomal enzyme **HMG CoA reductase** (mevalonate:$NADP^+$ oxidoreductase) that has an absolute requirement for NADPH as the reductant. This reductive step (1) consumes two molecules of NADPH, (2) results in hydrolysis of the thioester bond of HMG CoA, and (3) generates a primary alcohol residue in mevalonate. This reduction reaction is irreversible and produces (*R*)-(+)mevalonate, which contains six carbon atoms. HMG CoA reductase catalyzes the rate-limiting reaction in the pathway of cholesterol biosynthesis. HMG CoA reductase is an intrinsic membrane protein of the endoplasmic reticulum whose carboxyl terminus extends into the cytosol and carries the enzyme's active site.

The cholesterol pool of the body is derived from absorption of dietary cholesterol and biosynthesis primarily in liver and intestine. When the amount of dietary cholesterol is reduced, cholesterol synthesis is increased in liver and intestine to satisfy the needs of other tissues and organs. Cholesterol synthesized *de novo* is transported from liver and intestine to peripheral tissues in the form of lipoproteins. These are the only tissues that manufacture **apolipoprotein B,** the protein component of cholesterol transport proteins LDL and VLDL. Most apolipoprotein B is secreted into the circulation as VLDL, which is converted into LDL by removal of triacylglycerol and apolipoprotein C components, probably in

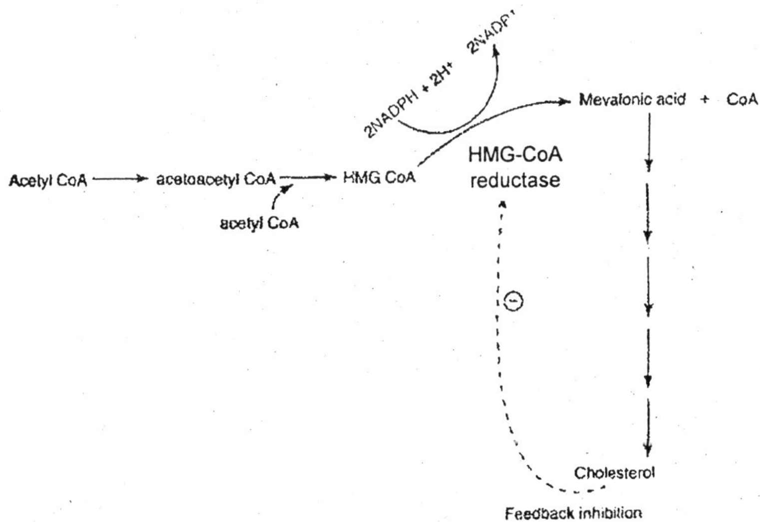

Summary of the pathway of cholesterol synthesis indicating feedback inhibition of HMG CoA reductase by cholesterol.

plasma and liver. When the quantity of dietary cholesterol increases, cholesterol synthesis in liver and intestine is almost totally suppressed. Thus the rate of *de novo* cholesterol synthesis is inversely related to the amount of dietary cholesterol taken up by the body.

In specialized tissues such as the adrenal gland and ovary, the cholesterol derived from LDL serves as a precursor to the steroid hormones made by these organs, such as cortisol and estradiol, respectively. In liver, cholesterol extracted from LDL and HDL is converted into bile salts that function in intestinal fat digestion.

Plasma cholesterol is in a dynamic state, entering the blood complexed with lipoproteins that keep the lipid in solution and leaving the blood as tissues take up cholesterol from these lipoproteins. Plasma lipoproteins contain free cholesterol and cholesterol esterified to a long-chain fatty acid. From 70% to 75% of plasma cholesterol is esterified to long-chain fatty acids. It is the free, unesterified form of cholesterol that exchanges readily between different lipoproteins and the plasma membrane of cells.

The mechanism for entry of cholesterol into liver cells from the three types of plasma lipoprotein is quite different. While the metabolism of chylomicrons and LDL has been quite well defined, that of HDL is just beginning to be understood. Chylomicrons that have had their triacylglycerol content reduced by plasma lipoprotein lipase become chylomicron remnants, which are rich in dietary cholesterol (free and esterified) and in fat-soluble vitamins. They are taken up by receptor-mediated endocytosis into liver cells. as is LDL.

High-density lipoproteins and the enzyme **lecithin:cholesterol acyltransferase** (LCAT) play important roles in the elimination of cholesterol from the body. LCAT catalyzes the freely reversible reaction, which transfers the fatty acid in the *sn*-2 position of phosphatidylcholine to the 3-hydroxyl of cholesterol. LCAT is a plasma enzyme produced mainly by liver. The actual substrate for LCAT is cholesterol contained in HDL. The LCAT-HDL system functions to protect cells, especially their plasma membrane, from the damaging effects of excessive amounts of free cholesterol. Cholesterol ester generated in the LCAT reaction diffuses into the core of the HDL particle where it is then transported from the tissues and plasma to liver, the latter being the only organ capable of metabolizing and excreting cholesterol. Thus, by this mechanism, referred to as the reverse transport of cholesterol, LCAT acting on HDL provides a vehicle for transporting cholesterol from peripheral tissues to the liver.

Many authorities recommend screening asymptomatic individuals by measuring plasma cholesterol. A level less than 200 mg% is considered desirable, and a level over 240 mg% requires lipoprotein analysis, especially determination of LDL cholesterol. Reduction of LDL cholesterol depends on dietary restriction of cholesterol to less than 300 mg day^{-1}, of calories to attain ideal body weight, and of total fat intake to less than 30% of total calories. Approximately two-thirds of the fat should be mono- or polyunsaturated. The second line of therapy is with drugs. Cholestyramine and colestipol are bile salt-binding drugs that promote excretion of bile salts in the stool. This in turn increases the rate of hepatic bile salt synthesis and of LDL uptake by the liver. Lovastatin is an inhibitor of HMG CoA reductase. Since this enzyme is limiting for cholesterol synthesis, lovastatin decreases endogenous synthesis of cholesterol and stimulates uptake and LDL via the LDL receptor. The combination of lovastatin and cholestyramine is sometimes used for severe hyperlipidemia.

6.7 Biosynthesis of Sphingosine

Sphingolipids are complex lipids whose core structure is provided by the long-chain amino alcohol **sphingosine** (4-sphingenine or *trans*-1,3-dihydroxy-2-amino-4-octadecene). Sphingosine has two asymmetric carbon atoms (C-2 and C-3); of the four possible optical isomers, naturally occurring sphingosine is of the D-erythro form. The double bond of sphingosine has the trans configuration. The primary alcohol group at C-1 is a nucleophilic center that forms covalent bonds with sugars to form glycosphingolipids and phosphocholine to form sphingomyelin. The amino group at C-2 always bears a long-chain fatty acid in amide linkage.

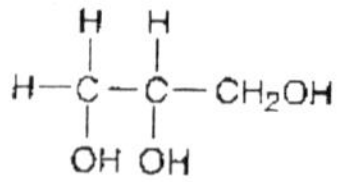

$$H-\overset{\displaystyle H}{\underset{\displaystyle OH}{C}}-\overset{\displaystyle H}{\underset{\displaystyle OH}{C}}-CH_2OH$$

Glycerol

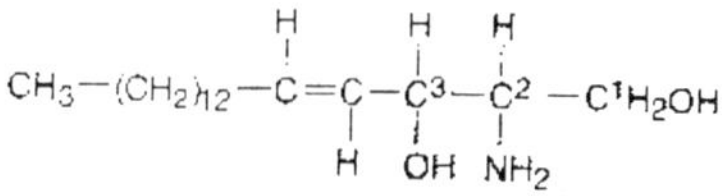

$$CH_3-(CH_2)_{12}-\overset{\displaystyle H}{C}=\underset{\displaystyle H}{C}-\overset{\displaystyle H}{\underset{\displaystyle OH}{C^3}}-\overset{\displaystyle H}{\underset{\displaystyle NH_2}{C^2}}-C^1H_2OH$$

Sphingosine

Comparison of the structures of glycerol and sphingosine (*trans*-1, 3,dihydroxy-2-amino-4-octadecene).

The secondary alcohol at C-3 is always free. It is useful to appreciate the structural similarity of a part of the sphingosine molecule to the glycerol moiety of the acylglycerols.

Sphingolipids are present in blood and nearly all body tissues. The highest concentrations are found in the white matter of the central nervous system. Various sphingolipids are components of the plasma membrane of practically all cells.

Sphingosine is synthesized by way of **sphinganine** (dihydrosphingosine) in two steps from the precursors serine and palmitoyl CoA. Serine is the source of C-1, C-2, and the amino group of sphingosine, while palmitic acid provides the remaining carbon atoms. Condensation of serine and palmitoyl CoA is catalyzed by a pyridoxal phosphate-dependent enzyme, serine palmitoyltransferase. The driving force for the reaction is provided by both cleavage of the reactive, high-energy thioester bond of palmitoyl CoA and the release of CO_2 from serine. The next step involves the reduction of the carbonyl group in 3-keto-dihydrosphingosine with reducing equivalents being derived from NADPH to produce sphinganine. The insertion of the double bond into sphinganine to produce sphingosine occurs at the level of ceramide.

Sphingosine does not occur naturally. The core structure of the natural sphingolipids is **ceramide,** a long-chain fatty acid amide

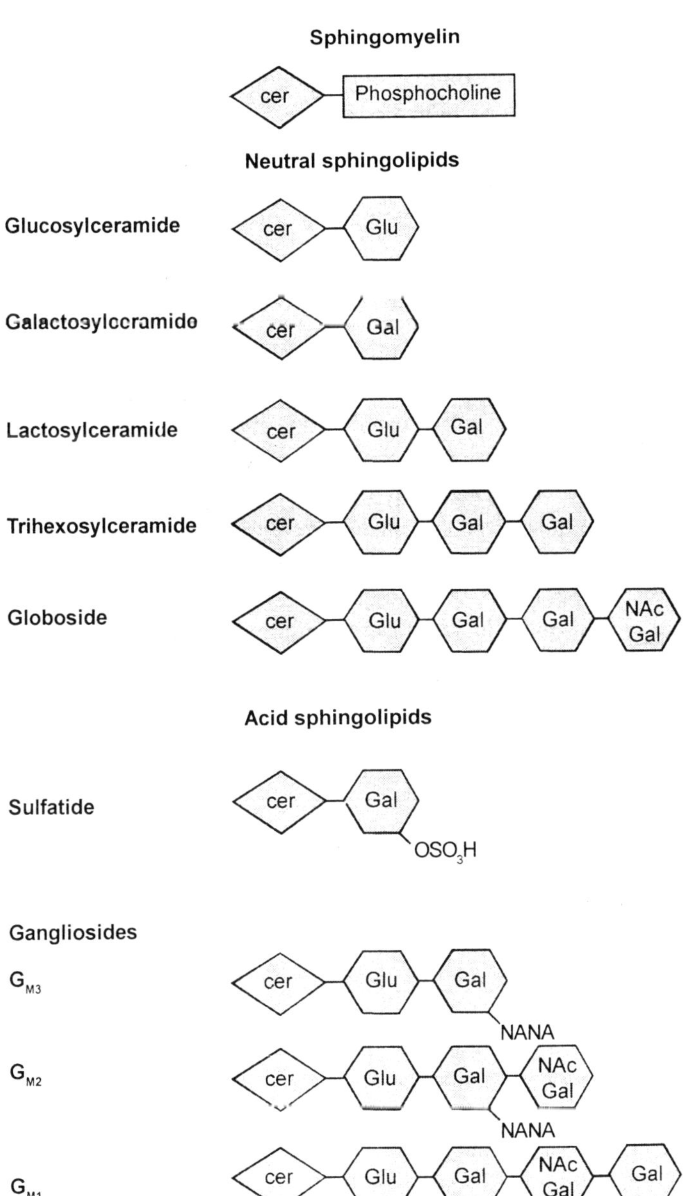

Structures of some common sphingolipids in diagrammatic form.

derivative of sphingosine. The long-chain fatty acid is attached to the 2-amino group of sphingosine through an amide bond. Most often the acyl group is **behenic acid,** a saturated C22 fatty acid, but other long-chain acyl groups can be used. There are two long-chain hydrocarbon domains in the ceramide molecule; these hydrophobic, regions are responsible for the lipid character of sphingolipids.

6.8 Sphingolipid Catabolism

Sphingolipids are normal degraded within lysosomes of phagocytic cells, particularly the histiocytes or macrophages of the reticuloendothelial system located primarily in liver, spleen, and bone marrow. Degradation of the sphingolipids by visceral organs begins with the engulfment of the membranes of white cells and erythrocytes that are rich in lactosylceramide (Cer-Glc-Gal) and hematoside (Cer-Glc-Gal-NANA). In the brain, the majority of the cerebroside-type lipids are gangliosides. Particularly during the neonatal period, ganglioside turnover in the central nervous system is extensive so that glycosphingolipids are rapidly being broken down and resynthesized. Note that among the various sphingolipids that comprise this pathway, there occurs a sulfate ester (in sulfolipid or sulfogalactolipid); *N*-acetylneuraminic acid groups (in the gangliosides); an α-linked galactose residue (in ceramide trihexoside); several *β*-galactosides (in galactocerebroside and G_{M1}); the ganglioside G_{M2}, which terminates in a *β*-linked *N*-acetyl-galactosamine unit; and glucocerebroside, which is composed of a single glucose residue attached to ceramide through a *β* linkage. The phosphodiester bond in sphingomyelin is broken to produce ceramide, which in turn is converted in sphingosine by the cleavage of an amide bond to a long-chain fatty acid. This overall pathway of sphingolipid catabolism is composed of a series of enzymes that cleave specific bond in the compounds including *α*- and *β*-galactosidases, a *β*-glucosidase, a neuraminidase, hexosaminidase, a sphingomyelin-specific phosphodiesterase (sphingomyelinase), a sulfate esterase) (sulfatase), and a ceramide-specific amidase (ceramidase). The important features of the sphingolipid catabolic pathway are as follows: (1) all the reactions take place within the

Sphingolipid Storage Diseases of Humans

Disorder	*Principal Signs and Symptoms*	*Principal Storage Substance*	*Enzyme Deficiency*
1. Tay-Sachs disease	Mental retardation, blindness, cherry red spot on macula, death between second and third year.	Ganglioside G_{M2}	Hexosaminidase A
2. Gaucher's disease	Liver and spleen enlargement, erosion of long bones and pelvis, mental retardation in infantile form only	Glucocerebroside	Glucocerebrosidase
3. Fabry's disease	Skin rash, kidney failure, pains in lower extremities	Ceramide trihexoside	α-Galactosidase A
4. Niemann-Pick disease	Liver and spleen enlargement, mental retardation.	Sphingomyelin	Sphingomyelinase
5. Globoid leukodystrophy (Krabbe's disease)	Mental retardation, absence of myelin	Galactocerebroside	Galactocerebrosidase
6. Metachromatic leukodystrophy	Mental retardation, nerves stain yellowish brown with cresyl violet dye (metachromasia)	Sulfatide	Arylsulfatase A
7. Generalized gangliosidosis	Mental retardation, liver enlargement, skeletal involvement	Ganglioside G_{M1}	G_{M1} ganglioside: β-galactosidase
8. Sandhoff-Jatzkewitz disease	Same as 1; disease has more rapidly progressing course	G_{M2} ganglioside, globoside	Hexosaminidase A and B
9. Fucosidosis	Cerebral degeneration, muscle spasticity, thick skin.	Pentahexosylfucoglycolipid	α-L-Fucosidase

lysosome; that is, the enzymes of the pathway are contained in lysosomes; (2) the enzymes are hydrolases; therefore one of the substrates in each reaction is water; (3) the pH optimum of each of the hydrolases is in the acid range, pH 3.5-5.5; (4) most of the enzymes are relatively stable and occur as isoenzymes; for example, **hexosaminidase** occurs in two forms: hexosaminidase A (HexA) and hexosaminidase B (HexB); (5) the hydrolases that comprise the sphingolipid pathway are glycoproteins and often occur firmly bound to the lysosomal membrane; and (6) the pathway is composed of intermediates that differ by only one sugar molecule, a sulfate group, or a fatty acid residue. The substrates are converted to products by the sequential, stepwise removal of constituents such as sugars and sulfate, by hydrolytic, irreversible reactions.

In most cases, sphingolipid catabolism functions smoothly, and all of the various complex glycosphingolipids and sphingomyelin are degraded to the level of their basic building blocks, namely, sugars, sulfate, fatty acid, phosphocholine, and sphingosine. However, when the activity of one of the hydrolytic enzymes is markedly reduced due to a genetic error, then the substrate for the defective or missing enzyme accumulates and is deposited within the lysosomes of the tissue responsible for the catabolism of that sphingolipid. For most of the reactions, patients have been identified who lack the enzyme that normally catalyzes that reaction. These disorders are called as sphingolipidoses.

We can generalize about some of the common features of **lipid storage diseases:** (1) usually only a single sphingolipid accumulates in the involved organs; (2) the ceramide portion is common to the various storage lipids; (3) the rate of biosynthesis of the accumulating lipid is normal; (4) a catabolic enzyme is missing in each of these disorders; and (5) the extent of the enzyme deficiency is the same in all tissues.

Diagnosis of a given sphingolipidosis can be made from a biopsy of the involved organ, usually bone marrow, liver, or brain, on morphologic grounds on the basis of the highly characteristic appearance of the storage lipid within lysosomes. Assay of enzyme activity is used to confirm the diagnosis of a particular lipid storage disease. Of great practical value is the fact that, for most of the diseases, peripheral leukocytes, cultured skin fibroblasts, and chorionic villi express the relevant enzyme deficiency and can be used as a source of enzyme for diagnostic purposes. In some cases (e.g. Tay-Sachs disease) serum and even tears are a source of enzyme for the diagnosis of a lipid storage disorder. Sphin-golipidoses, for the most part, are autosomal recessive, with the disease occurring only in homozygotes with a defect in both allelles. Enzyme assays can identify carriers or heterozygotes.

7

Carbohydrate Metabolism

7.1 Introduction

Carbohydrates produced during photosynthesis are utilized as a source of energy and as precursors for the synthesis of many structural and metabolic components. These requirements are satisfied in mammalian and human cells primarily by **D-glucose,** obtained from the diet or the storage compound, **glycogen**. This glucose is degraded by a metabolic pathway called **glycolysis**.

Glucose is the major form in which carbohydrate absorbed from the intestinal tract is presented to cells of the body. Glucose is the only fuel used to any significant extent by a few specialized cells and the major fuel used by the brain. Indeed, glucose is so important to these specialized cells and the brain that several of the major tissues of the body work together to ensure a continuous supply of this essential substrate. Glucose metabolism is defective in two very common metabolic diseases, obesity and diabetes, which contribute in development of a number of major medical problems, including atherosclerosis, hypertension, small vessel disease, kidney disease, and blindness.

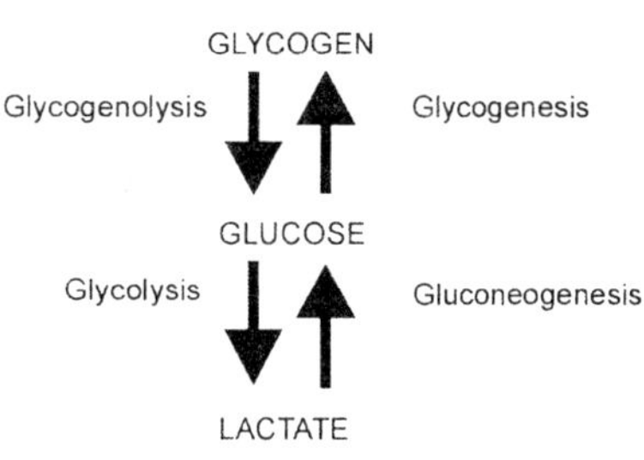

Relationship of glucose to the major pathways of carbohydrate metabolism.

The discussion begins with **glycolysis,** a pathway used by all cells of the body to extract part of the chemical energy inherent in the glucose molecule. This pathway also converts glucose to pyruvate and sets the stage for complete oxidation of glucose to CO_2 and H_2O. The *de novo* synthesis of glucose, that is, gluconeogenesis, is a function of the liver and kidneys and can conveniently be discussed following glycolysis because it makes use of some of the same enzymes used in the glycolytic pathway, although the reactions catalyzed are in the opposite direction. In contrast to glycolysis, which produces ATP, gluconeogenesis requries ATP and is therefore an energy-requiring process. The consequence is that only some of the enzyme-catalyzed steps can be common to both the glycolytic and gluconeogenic pathways. Indeed, additional enzymes including some in mitochondria become involved to make the overall process of gluconeogenesis exergonic. Regulation of the rate-limiting and key enzyme-catalyzed steps will be stressed throughout the chapter.

7.2 Glycolysis

The glycolytic pathway represents an ancient process, possessed by all cells of the human body, in which anaerobic degradation of glucose to lactate occurs. This is one example of anaerobic **fermentation,** a term used to refer to pathways by which organisms extract chemical energy from high-energy fuels in the absence of molecular oxygen. For many tissues glycolysis is an emergency energy-yielding pathway, capable of yielding 2 mol of ATP from 1 mol of glucose in the absence of molecular oxygen. Thus when the oxygen supply to a tissue is shut off, ATP levels can still be maintained by glycolysis for at least a short period of time. Many examples could be given, but the capacity to use glycolysis as a source of energy is particularly important to the human being at birth. With the exception of the brain, circulation of blood decreases to most parts of the body of the neonate during delivery. The brain is not normally deprived of oxygen during delivery, but other tissues must depend on glycolysis for their supply of ATP until circulation returns to normal and oxygen becomes available again. This conserves oxygen for use by the

brain, illustrating one of many mechanisms that have evolved to assure survival of brain tissue in times of stress. Glycolysis sets the stage for aerobic oxidation of carbohydrate. Oxygen is not necessary for glycolysis, and the presence of oxygen can indirectly suppress glycolysis, a phenomenon called the **Pasteur effect** that is considered later. Nevertheless, glycolysis can and does occur in cells with an abundant supply of molecular oxygen. Provided cells also contain mitochondria, the end product of glycolysis in the presence of oxygen is pyruvate rather than lactate. Pyruvate can then be completely oxidized to CO_2 and H_2O by enzymes housed within the mitochondria.

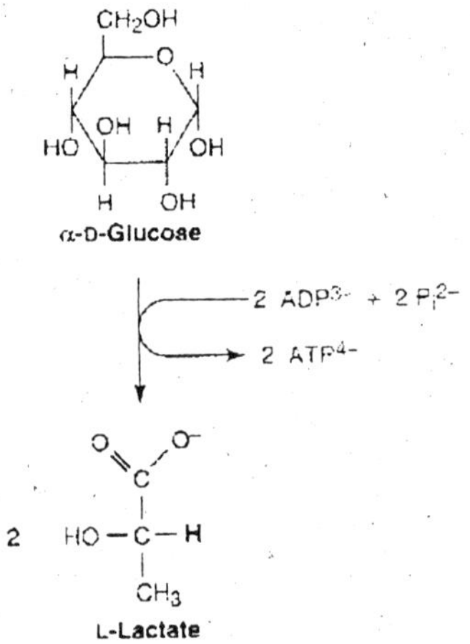

α-D-Glucose

$2\ ADP^{3-} + 2\ P_i^{2-}$ → $2\ ATP^{4-}$

2 L-Lactate

Overall balanced equation for the sum of the reactions of the glycolytic pathway.

Glycolysis can conveniently be pictured as occurring in three major stages.

Priming stage:

D-Glucose + 2 ATP^{4-} → D-fructose 1,6-bisphosphate^{4-} + 2 ADP^{3-} + 2 H^+

Splitting stage:

D-Fructose 1,6-bisphosphate^{4-} 2 D-glyceraldehyde 3-phosphate^{2-}

Oxidoreduction-phosphorylation stage:

$$2\ \text{D-Glyceraldehyde 3-phosphate}^{2-} + 4\ ADP^{3-} + 2\ P_i^{2-} + 2\ H^+ \rightarrow 2\ \text{L-lactate}^- + 4\ ATP^{4-}$$

Sum:

$$\text{D-Glucose} + 2\ ADP^{3-} + 2\ P_i^{2-} \rightarrow 2\ \text{L-lactate}^- + 2\ ATP^{4-}$$

Priming stage involves input of two molecules of ATP to convert glucose into a molecule of fructose 1,6-bisphosphate. ATP is therefore "invest" in the priming stage of glycolysis. However, ATP beyond this investment is gained from the glycolytic process. The **splitting stage** "splits" the six-carbon molecule fructose 1,6-bisphosphate into two molecules of glyceraldehyde 3-phosphate. In the **oxidoreduction-phosphorylation stage** two molecules of glyceraldehyde 3-phosphate are converted into two molecules of lactate with the production of four molecules of ATP. The overall process of glycolysis generates two molecules of lactate and two molecules of ATP at the expense of one molecule of glucose.

Hexokinase catalyzes the first step of glycolysis. Although this reaction consumes ATP, it gets glycolysis off to a good start by trapping glucose as glucose 6-phosphate (G6P) within the cytosol of the cell where all of the glycolytic enzymes are located. Phosphate esters are charged hydrophilic compounds that do not readily penetrate cell membranes. The phosphorylation of glucose with ATP is a thermodynamically favourable reaction, requiring the use of one high-energy phosphate bond. It is irreversible under cellular conditions. It is not, however, a way to synthesize ATP or to hydrolyze G6P to give glucose by the reverse reaction. Hydrolysis of G6P is accomplished by a different reaction, catalyzed by **glucose 6-phosphatase:**

$$\text{Glucose 6-phosphate}^{2-} + H_2O \rightarrow \text{glucose} + P_i^{2-}$$

This reaction is thermodynamically favourable in the direction written and cannot be used in cells for the synthesis of G6P from glucose. (A common mistake is to note that ATP and ADP are involved in the hexokinase reaction but not to note that they are not involved in the glucose 6-phosphatase reaction.) Glucose 6-phosphatase is an important enzyme in liver, functioning to produce free glucose from G6P in the last step of both gluconeogenesis and glycogenolysis; it has not role in glycolysis.

The next reaction is a readily reversible step of the glycolytic pathway, catalyzed by **phosphoglucose isomerase**. This step is not subject to regulation and, since it is readily reversible, functions in both glycolysis and gluconeogenesis.

6-Phosphofructo-1-kinase (or phosphofructokinase-1) catalyzes the next reaction, at ATP-dependent phosphorylation of fructose 6-phosphate (F6P) to give fructose 1,6-bisphosphate (FBP). This is a favorite enzyme of many students of biochemistry, being subject to regulation by several effectors and often considered the rate-limiting enzyme of the glycolytic pathway. The reaction is irreversible under intracellulor conditions.

Fructose 1,6-bisphosphate **aldolase** catalyzes the cleavage of frusctose 1,6-bisphosphate into a molecule each of dihydroxyacetone phosphate and glyceraldehyde 3-phosphate (GAP). This is a reversible reaction, the enzyme being called aldolase because the overall reaction is a variant of an aldol cleavage in one direction and an aldol condensation in the other. **Triose phosphate iso-merase** then catalyzes the reversible interconversion of dihydroxyacetone phosphate and GAP to complete the splitting stage of glycolysis. With the transformation of dihydroxyacetone phosphate (DHAP) into GAP, one molecule of glucose is converted into two molecules of GAP.

The first reaction of the last stage of glycolysis is catalyzed by **glyceraldehyde-3-phosphate dehydrogenase**. This reaction is of considerable interest because of what is accomplished in a single enzyme-catalyzed step. An aldehyde (glyceraldehyde 3-phosphate) is oxidized to a carboxylic acid with the reduction of NAD^+ to NADH. In addition to NADH, the reaction produces 1,3-bis-phosphoglycerate, a mixed anhydride of a carboxylic acid and phosphoric acid. 1,3-Bisphosphoglycerate has a large negative free energy of hydrolysis, enabling it to participate in a subsequent reaction that yields ATP. The overall reaction catalyzed by glyceraldehyde-3-phosphate dehydrogenase can be visualized as the coupling of a very favourable exergonic reaction with an unfavourable endergonic reaction. The exergonic reaction can be thought of as being composed of a half-reaction in which an aldehyde is oxidized to a carboxylic acid, which is then coupled with a half-reaction in which NAD⁺ is reduced to NADH:

$$R-\overset{\overset{\displaystyle O}{\|}}{C}H + H_2O \rightarrow R-\overset{\overset{\displaystyle O}{\|}}{C}OH + 2H^+ + 2e^-$$

$$NAD^+ + 2H^+ + 2e^- \rightarrow NADH + H^+$$

The overall reaction (sum of the half-reactions) is quite exergonic, with the aldehyde being oxidized to a carboxylic acid and NAD⁺ being reduced to NADH:

$$R-\overset{\overset{\displaystyle O}{\|}}{C}H + NAD^+ + H_2O \rightarrow R-\overset{\overset{\displaystyle O}{\|}}{C}OH + NADH + H^+, \Delta G^{\circ\prime} = -10.3\,\text{kcal mol}^{-1}$$

The endergonic component of the reaction corresponds to the formation of a mixed anhydride between the carboxylic acid and phosphoric acid:

$$R-\overset{\overset{\displaystyle O}{\|}}{C}OH + P_i^{2-} \rightarrow R-\overset{\overset{\displaystyle O}{\|}}{C}-OPO_3^{2-} + H_2O, \Delta G^{\circ\prime} = +11.8\text{ kcal mol}^{-1}$$

The overall reaction involves coupling of the endergonic and exergonic components to give an overall standard free-energy change of +1.5 kcal mol^{-1}.

$$\text{Sum: } R-\overset{\overset{\displaystyle O}{\|}}{C}H + NAD^+ + P_i^{2-} \rightarrow$$

$$R-\overset{\overset{\displaystyle O}{\|}}{C}OPO_3^{2-} + NADH + H^+, \Delta G^{\circ\prime} = +1.5\text{ kcal mol}^{-1}$$

7.3 The Regulatory Enzymes of the Glycolytic Pathway

The regulatory enzymes of the glycolytic pathway are hexokinase, 6-phosphofructo-1-kinase and pyruvate kinase.

A regulatory enzyme can often be identified by determining whether the concentrations of the substrates and products within a cell indicate that the reaction catalyzed by the enzyme is close to equilibrium.

An enzyme that is not subject to regulation will catalyze a "**near-equilibium reaction,**" whereas a regulatory enzyme will catalyze a "**nonequilibrium reaction**" under intracellular conditions. This makes sense because flux through the regulated enzyme is restricted by controls imposed on the enzyme. A nonregulatory enzyme is so active that it readily brings its substrates and products so equilibrium concentrations. Whether an enzyme-catalyzed reaction is near equilibrium or nonequilibrium can be determined by comparing the established equilibrium constant for the reaction with the mass-action ratio as it exists within a cell. The **equilibrium constant** for the reaction $A + B \rightleftharpoons C + D$ is defined as

$$K_{eq} = \frac{[C]\,[D]}{[A]\,[B]}$$

where the brackets indicate the concentrations at equilibrium. The **mass-action ratio** is calculated in a similar manner, except that the steady-state (ss) concentrations of reactants and products within the cell are used in the equation:

$$\text{Mass-action ratio} = \frac{[C]_{SS}\ [D]_{SS}}{[A]_{SS}\ [B]_{SS}}$$

If the mass-action ratio is approximately equal to the K_{eq}, the enzyme is said to be active enough to catalyze a near-equilibrium reaction and the enzyme is not considered subject to regulation. When the mass-action ratio is considerably different from the K_{eq}, the enzyme is said to catalyze a nonequilibrium reaction and usually will be found subject to regulation by one or more mechanisms.

The reactions catalyzed by **glucokinase** (liver isoenzyme of hexokinase), 6-phosphofructo-1-kinase, and pyruvate kinase in the intact liver are considered far enough from equilibrium to indicate that these enzymes are "regulatory" in the tissue.

The reaction catalyzed by glucokinase is not at equilibrium under the intracellular conditions of liver cells. Part of the explanation lies in the rate restriction imposed by the high K_m of glucokinase for glucose and part is due to the inhibitory protein.

Yet another important factor is that the activity of glucokinase is opposed in liver by that of **glucose 6-phosphatase.** Like glucokinase, this enzyme has a high K_m (3 mM) with respect to the normal intracellular concentration (~0.2 mM) of its primary substrate, glucose 6-phosphate. Thus the flux through this step is almost directly proportional to the intracellular concentration of glucose 6-phosphate. The combined action of glucokinase and glucose 6-phosphatase constitutes a futile cycle; that is, the sum of their reactions is hydrolysis of ATP to give ADP and P_i without the performance of any work. When blood glucose concentrations are about 5 mM, the activity of glucokinase is almost exactly balanced by the opposing activity of glucose 6-phosphatase. The result is that no net flux occurs in either direction. This futile cycling between glucose and glucose 6-phosphate is wasteful of ATP but, combined with the process of gluconeogenesis, contributes significantly to the "buffering" action of the liver on blood glucose levels. Furthermore, it provides a mechanism for preventing glucokinase from tying up all of the P_i of the liver.

Fructose, a component of many vegetables, fruits, and sweeteners, promotes hepatic glucose utilization by an indirect mechanism. It is converted in liver to fructose 1-phosphate, which activates glucokinase activity by promoting dissociation of the inhibitory protein. This may be a factor in the adverse effects (e.g., **hypertriglyceridemia**) sometimes associated with excessive dietary fructose consumption.

Glucokinase is an inducible enzyme. Under various physiological conditions the amount of the enzyme protein increases or decreases. Induction of synthesis and repression of synthesis of an enzyme are relatively slow processes, usually requiring several hours before significant changes occur. **Insulin** increases the amount of glucokinase by promoting transcription of the glucokinase gene. An increase in blood glucose levels signals an increase in insulin release from the β cells of the pancreas. This results in an increase in blood insulin levels, which promotes transcription of the glucokinase gene and increases the amount of liver glucokinase enzyme protein. Thus the amount of glucokinase in liver reflects how much glucose is being delivered to the liver via the portal vein. In other words, a person consuming large meals rich in carbohydrate will have greater amounts of glucokinase in the liver than one who is not. The liver in which glucokinase has been induced can make a greater contribution to the lowering of elevated blood glucose levels. The absence of insulin makes the liver of the diabetic patient deficient in glucokinase, in spite of high blood glucose levels, and this is one of the reasons why the liver of the diabetic has less blood glucose "buffering" action.

7.4 Gluconeogenesis

Net synthesis of formation of glucose from various substrates is termed **gluconeogenesis.** This includes use of various amino acids, lactate, pyruvate, propionate, and glycerol, as sources of carbon for the pathway. Glucose is also synthesized from galactose and fructose. **Glycogenolysis,** that is, formation of glucose or glucose 6-phosphate from glycogen, should be differentiated from gluconeogeneis; glycogenolysis rerfers to

$$\text{Glycogen or (glucose)}_n \rightarrow \text{molecules of glucose}$$

and thus does not correspond to *de novo* synthesis of glucose, the hallmark of the process of gluconeogenesis.

The capacity to synthesize glucose is crucial for survival of humans and other animals. Blood glucose levels have to be maintained to support metabolism of tissues that use glucose as their primary substrate. These include brain, red blood cells, kidney medulla, lens, cornea, testis, and a number of other tissues. Gluconeogenesis enables the maintenance of blood glucose levels long after all dietary glucose has been absorbed and completely oxidized.

Two important cycles between tissues that involve gluconeogenesis are recognized. The **Cori cycle** and the alanine cycle depend on gluconeogenesis in liver followed by delivery of glucose and its use in a peripheral tissue. Both cycles provide a mechanism for continuously supplying tissues that require glucose as their primary energy source. The cycle are only functional between liver and tissues that do not completely oxidize glucose to CO_2 and H_2O. In order to participate in these cycles, peripheral tissues must release either alanine or lactate as the end product of glucose metabolism. The type of recycled three-carbon intermediate is the major difference between the Cori cycle and the alanine cycle, carbon returning to liver as lactate in the Cori cycle but as alanine in the alanine cycle. Another major difference is that NADH generated by glycolysis in the alanine cycle cannot be used to reduce pyruvate to lactate. In tissues that have mitochondria, electrons of NADH can be transported into the mitochondria by the malate-aspartate shuttle or the glycerol phosphate shuttle for the synthesis of ATP by oxidative phosphorylation:

$$\text{NADH} + \text{H}^+ + \frac{1}{2}\,\text{O}_2 + 3\ \text{ADP} + 3\text{P}_i \rightarrow \text{NAD}^+ + 3\ \text{ATP} + \text{H}_2\text{O}$$

or

$$\text{FADH}_2 + \frac{1}{2}\,\text{O}_2 + 2\ \text{ADP} + 2\ \text{P}_i \rightarrow \text{FAD} + 2\ \text{ATP} + \text{H}_2\text{O}$$

The consequence is that six to eight molecules of ATP can be formed per glucose molecule in peripheral tissues that participate in the alanine cycle. This stands in contrast to the Cori cycle in which only two molecules of ATP per molecule of glucose are produced. Overall stoichiometry for the Cori cycle is

$$6\ \text{ATP}_{\text{liver}} + 2\ (\text{ADP} + \text{P}_i)_{\text{red blood cells}} \rightarrow 6\ (\text{ADP} + \text{P}_i)_{\text{liver}} + 2\ \text{ATP}_{\text{red blood cells}}$$

Six molecules of ATP are needed in liver to provide the energy necessary for glucose synthesis. The alanine cycle also transfers the energy from liver to peripherial tissues and, because of the six to eight molecules of ATP produced per molecule of glucose, is an energetically more efficient cycle. Participation of alanine in the cycle presents liver with amino nitrogen, which must be disposed of as urea. Urea synthesis is expensive since four ATP molecules are consumed per urea molecule. The concurrent need for urea synthesis results in more ATP being needed per glucose molecule synthesized in liver. Overall stoichiometry for the alanine cycle is

$$10\ ATP_{liver} + 6\text{-}8\ (ADP + P_i)_{muscle} + O_{2\ muscle} \rightarrow 10\ (ADP + P_i)_{liver} + 6\text{-}8\ ATP_{muscle}$$

In contrast to the Cori cycle, oxygen and mitochondria are required in peripheral tissue for participation in the alanine cycle.

Gluconeogenesis from lactate is an ATP-requiring process with the overall equation of

$$2\ \text{L-Lactate}^{-} + 6\ ATP^{4-} \rightarrow \text{glucose} + 6\ ADP^{3-} + 6\ P_i^{2-} + 4\ H^{+}$$

Many enzymes of glycolysis are common to the gluconeogenic pathway. Additional reactions have to be involved because glycolysis produces 2 ATPs and gluconeogenesis requires 6 ATPs per molecule of glucose. Also, certain steps of glycolysis are irreversible under intracellular conditions and are replaced by irreversible steps of the gluconeogenic pathway. The initial step is conversion of lactate to pyruvate by **lactate dehydrogenase.** NADH is generated and is also needed for a subsequent step in the pathway. Pyruvate cannot be converted to phosphoenolpyruvate (PEP) by **pyruvate kinase** because the reaction is irreversible under intracellular conditions. Pyruvate is converted into the high-energy phosphate compound PEP by coupling of two reactions requiring high-energy phosphate compounds (an ATP and a GTP). The first is catalyzed by pyruvate carboxylase and the second by PEP carboxykinase.

7.5 Glucose Synthesis

All **amino acids** except **leucine** and **lysine** can supply carbon for net synthesis of glucose by gluconeogenesis. If catabolism of an amino acid can yield either net pyruvate or net oxaloacetate formation, then net glucose synthesis can occur from that amino acid. Oxaloacetate is an intermediate in gluconeogenesis and pyruvate is readily converted to oxaloacetate by action of pyruvate carboxylase. The abbreviated pathway shows where amino acid catabolism fits with the process of gluconeogenesis. Catabolism of amino acids feeds carbon into the tricarboxylic cycle at more than one point. As long as net synthesis of a TCA cycle intermediate occurs as a consequence of catabolism of a particular amino acid, net synthesis of oxaloacetate will follow. Reactions that lead to net synthesis of TCA cycle intermediates are called **anaplerotic reactions (anaplerosis)** and support gluconeogenesis because they provide for net synthesis of oxaloacetate. Reactions catalyzed by pyruvate carboxylase and glutamate dehydrogenase are good examples of anaplerotic reactions:

$$\text{Pyruvate}^- + \text{ATP}^{4-} + \text{HCO}_3^- \rightarrow \text{oxaloacetate}^{2-} + \text{ADP}^{3-} + \text{P}_i^{2-} + \text{H}^+$$

$$\text{Glutamate}^- + \text{NAD(P)}^+ \rightleftharpoons \alpha\text{-ketoglutarate}^{2-} + \text{NAD(P)H} + \text{NH}_4^+ + \text{H}^+$$

On the other hand, the reaction catalyzed by glutamate-oxaloacetate transaminase (α-ketoglutarate + aspartate $\rightleftharpoons$ glutamate + oxaloacetate) is not anaplerotic because net synthesis of a TCA cycle intermediate is not accomplished. An intermediate of the TCA cycle is utilized in the reaction.

Since gluconeogenesis from amino acids imposes a nitrogen load on liver, a close relationship exists between urea synthesis and glucose synthesis from amino acids. Two alanine molecules are shown being metabolized, one yielding NH_4^+ and the other aspartate, the primary substrates for the urea cycle. Aspartate leaves the mitochondrion and becomes part of the urea cycle after reacting with citrulline. Carbon of aspartate is released from the urea cycle as fumarate, which is then converted to malate by cytosolic fumarase. Both this malate and another malate existing from the mitochondria are converted to glucose by cytosolic enzymes of gluconeogenesis. A balance is achieved between reducing equivalents (NADH) generated and those required in the cytosol and mitosol.

Leucine and **lysine** are the only amino acids that cannot function as carbon sources for net synthesis of glucose. These amino acids are **ketogenic** but not **glucogenic.** All other amino acids are classified as glucogenic, or at least both glucogenic and ketogenic. Glucogenic amino acids give rise to net synthesis of either pyruvate

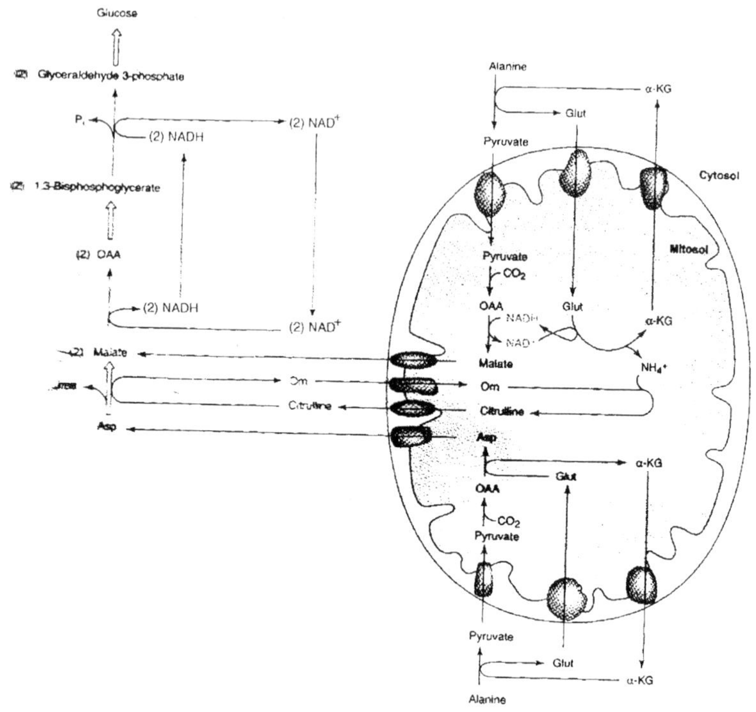

Pathway of gluconeogenesis from alanine and its relationship to urea synthesis.

or oxaloacetate, whereas amino acids that are both glucogenic and ketogenic also yield the ketone body acetoacetate, or acetyl CoA, which is readily converted into ketone bodies. Acetyl CoA is the end product of lysine metabolism, and acetoacetate and acetyl CoA are end products of leucine metabolism. No pathway exists for converting acetoacetate or acetyl CoA into pyruvate or oxaloacetate in humans and other animals. Acetyl CoA cannot be used for net synthesis of glucose because the reaction catalyzed by the **pyruvate dehydrogenase** comlex is irreversible.

$$\text{Pyruvate} + NAD^+ \text{ CoASH} \rightarrow \text{acetyl CoA} + NADH + CO_2$$

Humans consume considerable quantities of fructose in the form of sucrose hydrolyzed in the small bowel. In the liver, fructose is phosphorylated by a special ATP-linked kinase, yielding fructose 1-phosphate. A special aldolase then cleaves **fructose 1-phosphate** to yield one molecule of dihydroxyacetone phosphate and one of glyceraldehyde. The latter is reduced to glycerol and used by the same pathway given in the previous figure. Two molecules of dihydroxyacetone phosphate obtainable from one molecule of fructose can be converted to glucose by enzymes of gluconeogenesis or, alternatively, into pyruvate or lactate by the last stage of glycolysis. In analogy to glycolysis, conversion of fructose into lactate is called **fructolysis.**

The major energy source of spermatozoa is fructose, formed from glucose by cells of seminal vesicles. An NADPH-dependent reduction of glucose to sorbitol is followed by an NAD^+–dependent oxidation of sorbitol to fructose. Fructose is secreted from seminal vesicles in a fluid that becomes part of semen. Although the fructose concentration in human semen can exceed 10 mM, tissues that come in contact with semen utilize fructose poorly, allowing this substrate to be conserved to meet the energy demands of spermatozoa in their search for ova. Spermatozoa contain mitochondria and thus can metabolize fructose completely to CO_2 and H_2O by the combination of fructolysis and TCA cycle activity.

Milk sugar or **lactose** is an important source of **galactose** in the human diet. Glucose formation from galactose follows the pathway. **UDP-glucose** serves as a recycling intermediate in the overall process of converting galactose into glucose. Absence of the enzyme **galactose 1-phosphate uridylyltransferase** accounts for most cases of **galactosemia**.

Mannose is found in very limited quantities in our diet. It is phosphorylated by hexokinase and then converted into fructose 6-phosphate by mannose phosphate isomerase:

D-Mannose + ATP → D-mannose 6-phosphate + ADP

D-Mannose 6-phosphate → D-fructose 6 phosphate

The latter compound can then be used in either glycolysis or gluconeogenesis.

Pathway of glucose formation from fructose, along with the competing pathway of fructolysis, can be shown as follows.

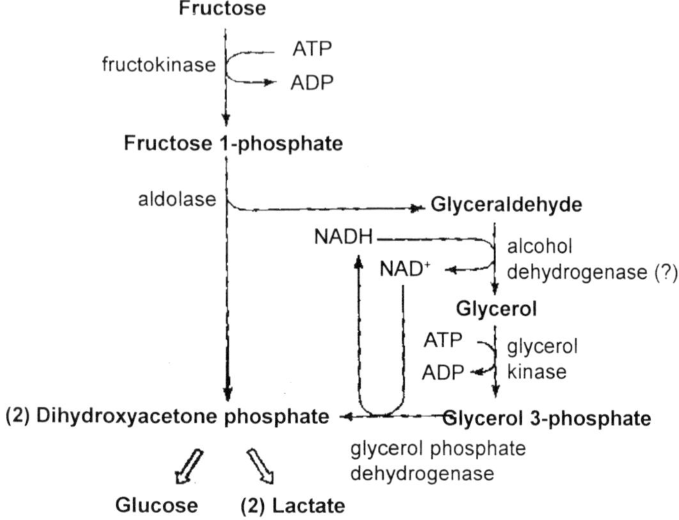

Hormonal control of gluconeogenesis is a matter of regulating the supply of fatty acids to liver and the enzymes of both the glycolytic and gluconeogenic pathways. Glucagon increases plasma fatty acids by promoting lipolysis in adipose tissue, an action opposed by insulin. The greater availability of fatty acids results in more fatty acid oxidation by liver, which promotes glucose synthesis. Insulin has the opposite effect. Glucagon and insulin also regulate gluconeogenesis by influcencing the state of phosphorylation of hepatic enzymes subject to covalent modification. As discussed previously, pyruvate kinase is active in the dephosphorylated mode and inactive in the phosphorylated mode. Glucagon activates adenylate cyclase to produce cAMP, which activates protein kinase A, which, in turn, phosphorylates and inactivates pyruvate kinase. Inactivation of this glycolytic enzyme stimulates the opposing pathway gluconeogenesis, by blocking the futile conversion of PEP to pyuvate. Glucagon also stimulates gluconeogenesis at the conversion of fructose 1,6-bisphosphate to fructose 6-phosphate by decreasing the concentration of fructose 2,6-bisphosphate in liver.

Fructose 2,6-bisphosphate is an allosteric activator of 6-phosphofructo-1-kinase and an allosteric inhibitor of fructose 1,6-bisphosphatase. Glucagon, again working via its second messenger cAMP, lowers fructose 2,6-bisphosphate levels by stimulating the phosphorylation of the bifunctional enzyme 6-phosphofructo-2-kinase/fructose 2,6-bisphosphatase. Phosphorylation of this enzyme inactivates the kinase activity that makes fructose 2,6-bisphosphate from F6P but activates the phosphatase activity that hydrolyzes fructose 2,6-bisphosphate back to F6P. The consequence is a glucagon-induced fall in fructose 2,6-bisphosphate levels, leading to a decrease in activity of 6-phosphofructo-1 kinase while fructose 1,6-bisphosphatase becomes more active. The overall effect is an increased conversion of FBP to F6P and a corresponding increase in the rate of gluconeogenesis. A resulting increase in fructose 6-phosphate may also favour gluconeogenesis by inhibition of glucokinase via an inhibitory protein. Insulin has effects opposite to those of glucagon by mechanisms not completely defined.

Glucagon and insulin also have long-term effects on hepatic glycolysis and gluconeogenesis by induction and repression of synthesis of key enzymes of the pathways. A high glucagon/insulin ratio in blood increases the enzymatic capacity for gluconeogenesis and decreases enzymatic capacity for glycolysis in liver. A low glucagon/insulin ratio has the opposite effects. The glucagon/insulin ratio increases when gluconeogenesis is needed and decreases when glucose is plentiful from the gastrointestinal tract. Glucagon signals induction of synthesis of greater quantities of PEP carboxykinase, fructose 1,6-bisphosphatase, glucose 6-phosphatase, and various aminotransferases. Binding of glucagon to its plasma membrane receptor increases cAMP, which activates protein kinase.

7.6 Glycogenolysis

Glycogenolysis refers to breakdown of glycogen to glucose or glucose 6-phosphate; and glycogenesis refers to synthesis of glycogen. These processes are of some importance in almost every tissue but especially in muscle and liver.

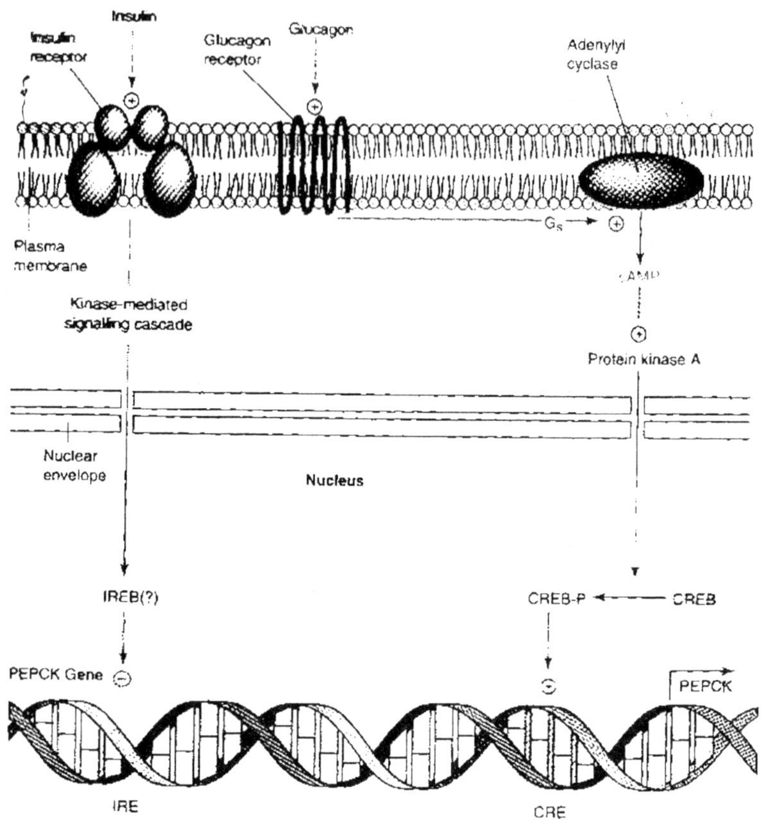

Glucagon promotes transcription of the gene that encodes PFP carboxykinase.

Muscle and liver **glycogen stores** serve completely different roles. Glycogen serves as a fuel reserve for the synthesis of ATP within mucle, whereas liver glycogen functions as a glucose reserve for the maintenance of blood glucose concentrations. Liver glycogen levels vary greatly in response to the intake of food, accumulating to high levels shortly after a meal and then decreasing slowly as it is mobilized to help maintain a nearly constant blood glucose level. Liver glycogen is called into play between meals and to a greater extent during the nocturnal fast. In both humans and the rat, the store of liver glycogen lasts somewhere between 12 and 24 h during fasting, depending greatly, of course, on whether the individual under consideration is caged or running wild.

Muscle glycogen is a source of ATP for increased muscular activity. Most of the glucose of glycogen is consumed within muscle cells without formation of free glucose as an intermediate. However, because of a special feature of glycogen catabolism to be discussed below, about 8% of muscle glucogen is converted into free glucose within the tissue. Some of this glycose may be released into the bloodstream, but most gets metabolized by glycolysis in muscle. Since muscle lacks glucose 6-phosphatase, and most free glucose formed during glycogen breakdown is further catabolized, muscle glycogen is not of quantitative importance in maintenance of blood glucose levels in the fasting state. Liver glycogen converted to glucose by glycogenolysis and glucose 6-phosphatase is of much greater importance as a source of blood glucose in the fasting state. Conversion of glucose to glycogen in muscle plays an important role in lowering blood glucose levels elevated by a high carbohydrate meal. Glycogenesis in liver contributes to the lowering of blood glucose but is of less importance than glycogen synthesis in muscle.

Exercise of a muscle triggers mobilization of muscle glycogen for formation of ATP. The yield of ATP and the fate of the carbon of glycogen depend on whether a "white" or "red" muscle is under consideration. Red muscle fibers are supplied with a rich blood flow, contain large amounts of myoglobin, and are packed with mito-chondria. Glycogen mobilized within these cells is converted into pyruvate, which, because of the availability of O_2 and mitochondria, can be converted into CO_2 and H_2O. In contrast, white muscle fibres have a poorer blood supply and fewer mitochondria. **Glycogenolysis** within this tissue supplies substrate for glycolysis, with the end product being primarily lactate. **White muscle fibers** have enormous capacity for glycogenolysis and glycolysis, much more than **red muscle fibers.** Since their glycogen stores are limited, however, muscles of this type can only function at full capacity for relatively short periods of time. Breast muscle

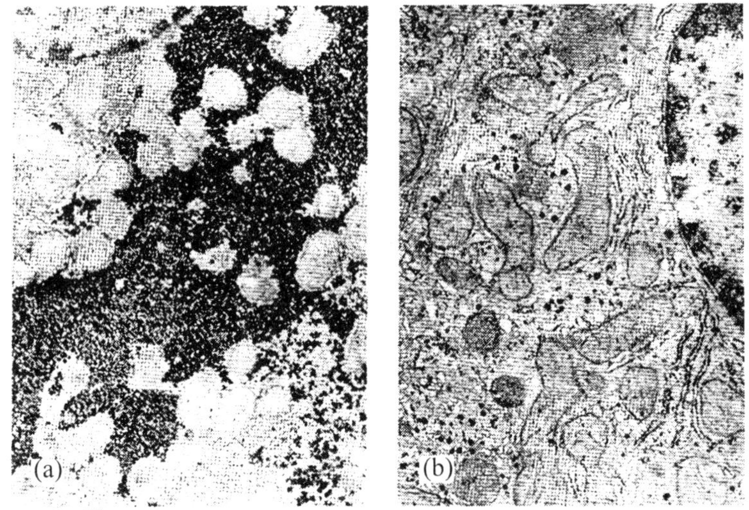

Electron micrographs showing glycogen granules (darkly stained material) in the liver of a well-fed rat (a) and the relative absence of such granules in the liver of a rat starved for 24 h (b).

and the heart of chicken are good examples of white and red muscles, respectively.

There are a number of well-characterized glycogen storage diseases, all due to inherited defects of one or more of the enzymes involved in the synthesis and degradation of glycogen. The liver is usually the tissue most affected, but heart and muscle glycogen metabolism can also be defective.

The most common glycogen storage disease, referred to as type I or von Gierke's disease, is caused by a deficiency of liver, intestinal mucosa, and kidney glucose 6-phosphatase. Thus diagnosis by small bowel biopsy is possible. Patients with this disease can be further subclassified into those lacking the glucose 6-phosphatase enzyme per se (type Ia) and those lacking the glucose 6-phosphatase translocase (type Ib). A genetic abnormalitly in glucose 6-phosphate hydrolysis occurs in only about 1 person in 200,000 and is transmitted as an autosomal recessive trait. Clinical manifestations include fasting hypoglycemia, lactic acidemia hyperlipidemia, and hyperuricemia with gouty arthritis. The fasting hypoglycemia is readily explained as a consequence of the glucose 6-phosphatase deficiency, the enzyme required to obtain glucose from liver glycogen and gluconeogenesis. The liver of these patients dose release some glucose by the action of the glycogen debranching enzyme. The lactic acidemia occurs because the liver cannot use lactate effectively for glucose synthesis. In addition, the liver inappropriately produces lactic acid in response to glucagon. This hormone should trigger glucose release without lactate production; however, the opposite occurs because of the lack of glucose 6-phosphatase. Hyperuricemia results from increased purine degradation in the liver, hyperlipidemia results because of increased availability of lactic acid for lipogenesis and lipid mobilization from the adipose tissue caused by high glucagon levels in response to hypoglycemia. The manifestations of von Gierke's disease can greatly be diminished by providing carbohydrate throughout the day to prevent hypoglycemia. During sleep this can be done by infusion of carbohydrate into the gut by a nasogastric tube.

The first reaction involved in **glycogen synthesis** is already familiar, being catalyzed by glucokinase in hepatic tissue and hexokinase in peripheral tissues.

$$\text{Glucose} + \text{ATP} \rightarrow \text{glucose 6-phosphate} + \text{ADP}$$

Phosphoglucomutase, discussed in relation to glycogen degradation, catalyzes a readily reversible reaction as follows:

$$\text{Glucose 6-phosphate} \rightarrow \text{glucose 1-phosphate}$$

A unique reaction found at the next step involves formation of UDP-glucose by action of **glucose 1-phosphate uridylyltransferase:**

$$\text{Glucose 1-phosphate} + \text{UTP} \rightarrow \text{UDP-glucose} + PP_i$$

This reaction generates UDP-glucose, sometimes called **"activated glucose"** because of its large negative free energy of hydrolysis, which is used to build the glycogen molecule. Formation of UDP-glucose is made energetically favourable and irreversible by hydrolysis of pyrophosphate by **pyrophosphatase:**

$$PP_i^{4-} + H_2O \rightarrow 2\ P_i^{2-}$$

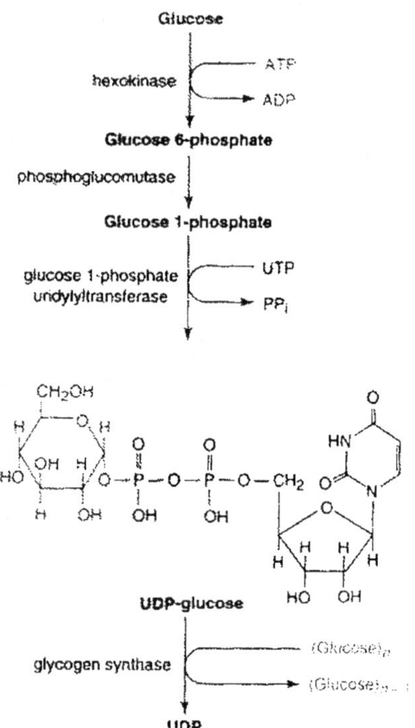

Pathway of glycogen synthesis.

Glycogen synthase catalyzes transfer of the activated glucosyl moiety of UDP-glucose to a glycogen molecule to form a new glycosidic bond between the hydroxyl group of C-1 of the activated sugar and C-4 of a glucosyl residue of the growing glycogen chain. The reducing end of glucose (C-1 of glucose is an aldehyde that can reduce other compounds) is always added to a nonreducing end of the glycogen chain. The glycogen molecule, regardless of its size, theoretically has only one free reducing end tucked away within its core. UDP formed as a product of glycogen synthase is converted back to UTP by action of **nucleoside diphosphate kinase:**

$$\text{UDP} + \text{ATP} \rightarrow \text{UTP} + \text{ADP}$$

Glycogen synthase creates chains of glucose molecules with α-1,4-glycosidic linkages, but does not form the α-1,6-glycosidic branches found in glycogen. Its action alone would only produce

α-amylose, a straight-chain polymer of glucose with α-1,4-glycosidic linkages. Once an amylose chain of at least 11 residues has been formed, a **"branching" enzyme** comes into play. Its name is **1,4-α-glucan branching enzyme** because it removes a block of about seven glucosyl residues from a growing chain and transfers it to another chain to produce an α-1,6 linkage. The new branch has to be introduced at least four glucosyl residues from an adjacent branch point. Thus the creation of the highly branched structure of glycogen requires the concerted efforts of glycogen synthase and branching enzyme.

Since **glycogen** is such a good fuel reserve, it is obvious why we synthesize and store glycogen in liver and muscle. But why not store our excess glucose calories entirely as fat instead of glycogen? The answer is at least threefold: (1) we do store **fat**, but fat cannot be mobilized nearly as rapidly in muscle as glycogen; (2) fat cannot be used as a source of energy in the absence of oxygen; and (3) fat cannot be converted to glucose to maintain blood glucose levels.

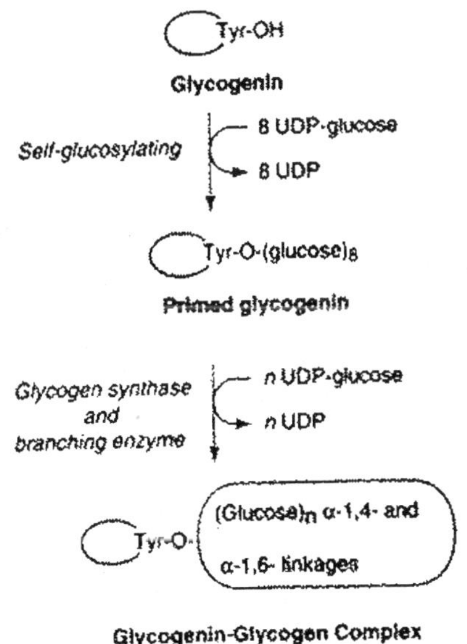

Glycogenin provides a primer for glycogen synthesis by glycogen synthase.

Why not just pump glucose into cells and store it as free glucose until needed? Why waste so much ATP making a polymer out of glucose? The problem is that glucose is osmotically active. It would cost ATP to "pump" glucose into a cell against a concentration gradient, and glucose would have to reach concentrations of 400 mM in liver cells to match the "glucose reserve" provided by the usual liver glycogen levels. Unless balanced by outward movement of some other osmotically active compound, accumulation of such concentrations of glucose would cause uptake of considerable water and osmotic lysis of the cell. Assuming the molecular mass of a glycogen molecule is of the order of 10^7 Da, 400 mM glucose is in effect stored at an intracellular glycogen concentration of 0.01 μM. Storage of glucose as glycogen therefore creates no osmotic pressure problem for the cell.

Like DNA synthesis, a primer is needed for glycogen synthesis. No template, however, is required. Glycogen itself is the usual **primer,** in that glycogen synthesis can take place by addition of glucosyl units to glycogen "core" molecules, which are almost invariably present in the cell. The outer regions of the glycogen molecule get removed and resynthesized more rapidly than the inner core. Glycogen within a cell is frequently sheared by the combined actions of glycogen phosphorylase and debranching enzyme but is seldom obliterated before glycogen synthase and branching enzyme rebuild the molecule. This begs the question why glycogen is a branched molecule with only one real beginning (the reducing end) and many branches terminating with nonreducing glucosyl units. The answer is that this gives numerous sites of attack for glycogen phosphorylase on a mature glycogen molecule and the same number of sites that function as primers for the addition of glucosyl units by glycogen synthase. If cells synthesized **α-amylose,** that is, an unbranched glucose polymer, there would only be one nonreducing end per molecule. This would surely make glycogen degradation and synthesis much slower. As it is, glycogen phosphorylase and glycogen synthase are found in tight association with glycogen granules in a cell. By taking up residence in the branches of the glycogen tree, both enzymes have ready access to a multitude of nonreducing sugars at the ends of the limbs.

But why is a primer needed for glycogen synthesis? It turns out to be impossible to initiate glycogen synthesis with simply a glucose molecule as the acceptor of an activated glucosyl residue from UDP-glucose. Glycogen synthase has a very low K_m for very large glycogen molecules and therefore readily adds glucosyl residues to make even larger glycogen molecules. However, the K_m gets larger and larger as the glycogen molecule gets smaller and smaller. This phenomenon is so pronounced that glucose, at its physiological concentration, could never function as a primer. This led for some time to the notion that glycogen must be immortal; that is, some glycogen must be handed down from one cell generation to the next in order for glycogen to be synthesized. However, it is now known that a polypeptide of 332 amino acids called **glycogenin** functions as a primer for glycogen synthesis. Glycogenin is a self-glucosylating enzyme that uses UDP-glucose to link glucose to one of its own tyrosine residues. Glycosylated glycogenin then serves as a primer for synthesis of glycogen. Alas, glycogen is not immortal.

Glycogen synthase has to be active for glycogen synthesis and inactive for glycogen degradation. The combination of the reactions catalyzed by glycogen synthase, glycogen phosphorylase, glucose 1-phosphate uridylyltransferase, and nucleoside diphosphate kinase adds up to a futile cycle with the overall equation ATP $\rightarrow$ ADP + P_i. Hence glycogen synthase needs to be turned off when glycogen phosphorylase is turned on, and vice versa.

Activation of **glycogen synthase** by glucose 6-phosphate, an allosteric effector, is probably of physiological significance under some circumstances. However, as with glycogen phosphorylase, this mode of control is integrated with regulation by covalent modification. Glycogen synthase exists in two forms. One is designated the D form because it is dependent on the presence of G6P for activity. The other is designated the I form because its activity is independent of the presence of G6P. The D form corresponds to the *b* or inactive form of the enzyme, the I form to the *a* or active form of the enzyme. Phosphorylation of glycogen synthase is catalyzed by several different kinases, which in turn are regulated by second messengers of hormone action, including cAMP, Ca^{2+}, **diacylglycerol,** and probably yet to be identified compounds. Each of the protein kinases is capable of catalyzing the phosphorylation and contributing to inactivation of glycogen synthase. Although glycogen synthase is a simple tetramer (α_4) of only one subunit type (mol wt 85,000 Da), it can be phosphorylated on at least nine different serine residues. Eleven different protein kinases have been identified that can phosphorylate glycogen synthase. This stands in striking contrast to glycogen phosphorylase, which is regulated by phosphorylation of one site by one specific kinase.

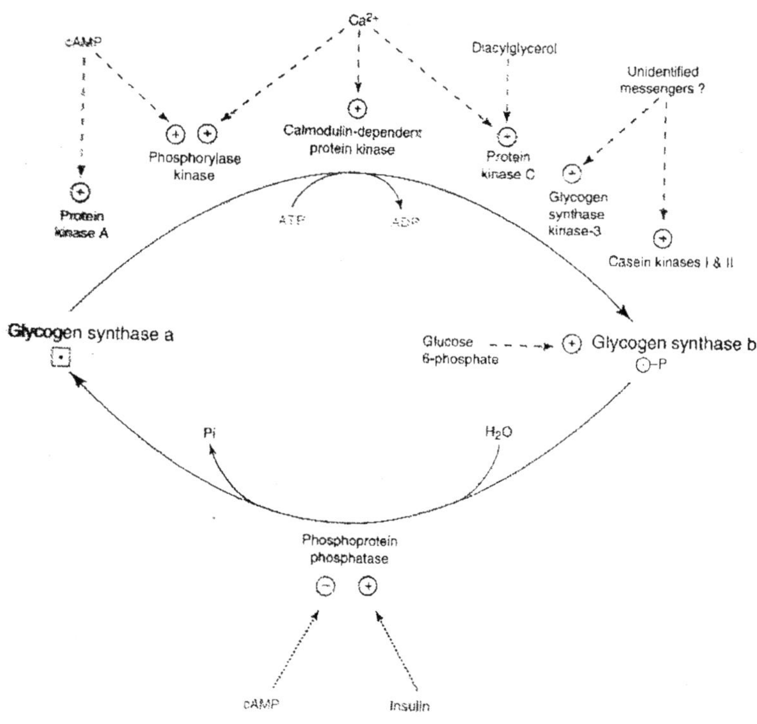

Regulation of glycogen synthase by covalent modification.

Cyclic AMP is an extremely important intracellular signal for reciprocally controlling glycogen synthase and glycogen phosphorylase. An increase in cAMP signals activation of glycogen phosphorylase and inactivation of glycogen synthase via activation of protein kinase A and inhibition of phosphoprotein phosphatase. Ca^{2+} likewise can influence the phosphorylation state of both enzymes and reciprocally regulate their activity via its effects on phosphorylase kinase. Two cAMP-independent, Ca^{2+}-activated protein kinases have been identified that also may have physiological significance. One of these is a calmodulin-dependent protein kinase and the other a Ca^{2+}-and phospholipid-dependent protein kinase (protein kinase C). Both enzymes phosphorylate glycogen synthase, but neither can phosphorylate glycogen phosphorylase. Protein kinase C requires phospholipid, diacylglycerol, and Ca^{2+} for full activity. There is considerable interest in protein kinase C because **tumor-promoting agents** called **phorbol esters** have been found to minic diacylglycerol as activators of this enzyme. Diacylglycerol is considered an important "second messenger" of hormone action, acting via protein kinase C to regulate numerous cellular process.

Glycogen synthase is also phosphorylated by glycogen synthase kinase-3, casein kinase 1, and casein kinase II. These kinases are not subject to regulation by cAMP or Ca^{2+}. It is likely, however, that special regulatory mechanisms exist to regulate these kinases. Herein may lie solutions to unsolved problems such as the mechanism of action of insulin and other hormones.

The phosphoprotein phosphatase that converts glycogen synthase *b* back to glycogen synthase *a* is regulated in a manner analogous to that described in the discussion of glycogen phosphorylase regulation. Cyclic AMP promotes inactivation whereas insulin promotes activation of glycogen synthase through opposite effects of phosphoprotein phosphatase activity.

7.7 Glucagon and Epinephrine

Glucagon is released from α cells of pancreas in response to low blood glucose levels. One of glucagon's primary jobs during periods of low food intake (fasting or starvation) is to mobilise **liver glycogen,** that is, stimulate glycogenolysis in order to ensure that adequate blood glucose is available to meet the needs of glucose-dependent tissues. Glucagon circulates in blood until it interacts with glucagon receptors such as those located on the plasma membrane of liver cells. Binding of glucagon to these receptors activates adenylate cyclase and triggers the cascades that result in activation of glycogen phosphorylase and inactivation of glycogen synthase by the mechanisms. Glucagon also inhibits glycolysis at the level of 6-phosphofructo-1-kinase and pyruvate kinase by the mechanisms. The net result of these effects of glucagon, all mediated by the second messenger cAMP and covalent modification, is a very rapid increase in blood glucose levels. Hyperglycemia might be expected but does not occur because less glucagon is released from the pancreas as blood glucose levels increase.

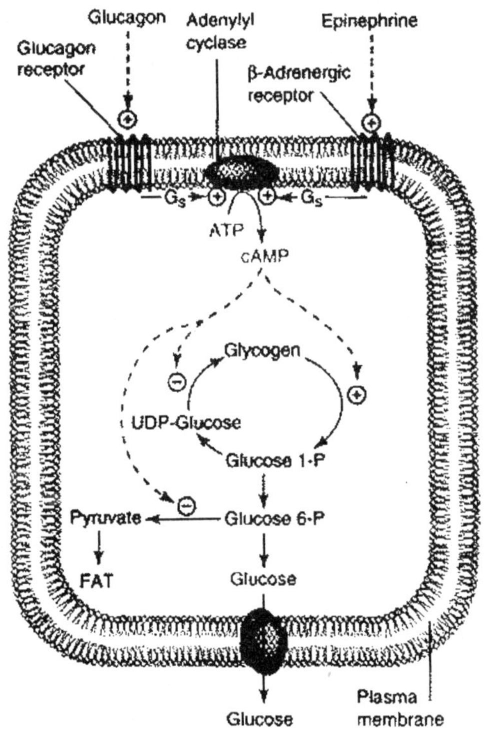

Cyclic AMP mediates the stimulation of glycogenolysis in liver by glucagon and *β* agonists (epinephrine).

Epinephrine is released into blood from chromaffin cells of the adrenal medulla in response to stress. This hormone is our "fright, flight, or fight" hormone, preparing the body for either combat or escape.

Epinephrine interacts directly with receptors in the plasma membrane of liver cells to activate **adenylate cyclase**. The resulting increase in cAMP has the same effect as that caused by glucagon, that is, activation of glycogenolysis and inhibition of glycogenesis and glycolysis to maximize the release of glucose from liver. The plasma membrane receptor for epinephrine, which is in communication with adenylate cyclase, is the β-adrenergic receptor. The plasma membrane of liver cells also has another binding protein for epinephrine, called the **α-adrenergic receptor.** Interaction of epinephrine with α-adrenergic receptors leads to formation of **inositol 1,4,5-trisphosphate** (IP_3) and **diacylglycerol**. These

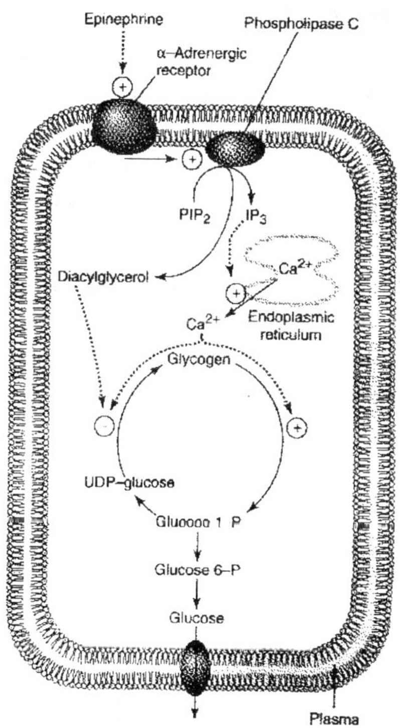

compounds are second messengers, produced in the plasma membrane by the action of a phospholipase C on phosphatidylinositol 4,5-bisphosphate. Inositol 1,4,5-trisphosphate stimulates the release of Ca^{2+} from the endoplasmic reticulum. As previously discussed, the increase in Ca^{2+} activates phosphorylase kinase, which in turn activates glycogen phosphorylase. Likewise Ca^{2+}-mediated activation of phosphorylase kinase, calmodulin-dependent protein kinase, and protein kinase C, as well as diacylglycerol-mediated activation of protein kinase C, may all be important for inactivation of glycogen synthase.

The consequences of epinephrine action is an increased release of glucose into the blood from the glycogen stored in liver. This makes more blood glucose available to tissues that are called upon to meet the challenge of the stressful situation that triggered the release of epinephrine from adrenal medulla.

7.8 Sugar Transformations

Formation of some sugar can occur directly, starting from glucose via modification reactions such as the conversion of G6P to fructose 6-phosphate by phosphoglucose isomerase in the glycolitic pathway. A similar aldose-ketose isomerization catalyzed by phosphomannose isomerase results in synthesis of mannose 6-phosphate.

Internal transfer of a phosphate group on the same sugar molecule from one hydroxyl group to another is a common modification. Glucose 1-phosphate, resulting from enzymatic phosphorolysis of glycogen, is converted to G6P by phosphoglucomutase. Galactose can be phosphorylated directly to galactose 1-phosphate by a galactokinase and mannose to mannose 6-phosphate by a mannokinase. Free fructose, an important dietary constituent, can be phosphorylated in the liver to fructose 1-phosphate by a special fructokinase. However, no mutase exists to interconvert fructose 1-phosphate and fructose 6-phosphate, nor can phosphofructokinase synthesize fructose 1,6-bisphosphate from fructose 1-phosphate. Rather, a fructose 1-phosphate aldolase cleaves fructose 1-phosphate to dihydroxyacetone phosphate

(DHAP), which enters the glycolytic pathway directly, and glyceraldehyde, which must first be reduced to glycerol, phosphorylated, and then reoxidized to DHAP. Lack of this aldolase leads to fructose intolerance.

Most other sugar transformation reactions require prior conversion into **nucleotide-linked sugars.** Formation of nucleoside diphosphate (NDP)-sugar involves the reaction of hexose 1-phosphate and nucleoside triphosphate (NTP), catalyzed by a pyrophosphorylase. While these reactions are readily reversible, *in vivo* pyrophosphate is rapidly hydrolyzed by pyrophosphatase, thereby driving the synthesis of nucleotide sugars.

Fructose may account for 30-60% of the total carbohydrate intake of mammals. It is predominantly metabolized by a specific fructose pathway. The first enzyme in this pathway, fructokinase, is deficient in essential fructosuria. This disorder is a benign asymptomatic metabolic anomaly, which appears to be inherited as an autosomal recessive. Following intake of fructose, blood levels and urinary fructose are unusually high; however, 90% of fructose is eventually metabolized. In contrast, hereditary fructose intolerance is characterized by severe hypoglycemia after ingestion of fructose. Prolonged ingestion in young children may lead to death. In this disorder fructose 1-phosphate aldolase is deficient, and fructose 1-phosphate accumulates intracellularly.

Epimerization is a common type of reaction in carbohydrate metabolism. Reversible conversion of glucose to galactose in animals occurs by epimerization of UDP-glucose to UDP-galactose, catalyzed by UDP-glucose epimerase. UDP-galactose is also an important intermediate in metabolism of free galactose, derived from the hydrolysis of lactose in the intestinal tract. Galactose is phosphorylated by galactokinase and ATP to yield galactose 1-phosphate. Then galactose 1-phosphate uridylyltransferase transforms galactose 1-phosphate into UDP-galactose by displacing glucose 1-phosphate from UDP-glucose. These reactions are summarized as follows:

$$\text{Galactose} + \text{ATP} \rightleftharpoons \text{galactose 1-phosphate} + \text{ADP}$$

UDP-glucose + galactose 1-phosphate ⇌ UDP galactose + glucose 1-phosphate

A hereditary disorder, galactosemia, results from the absence of this uridylyl-transferase.

A combination of these reactions allows an efficient transformation of dietary galactose into glucose 1-phosphate, which can then be further metabolized by previously described pathways. Alternatively, the 4-epimerase can operate in the reverse direction when UDP-galactose is needed for biosynthesis. Epimerization reactions are not exclusively restricted to nucleotide-linked sugars but also occur at the polymer level; D-glucuronic acid is epimerized to L-iduronic acid after incorporation into heparin and dermatan sulfate.

Reactions of galactose are of particular interest because in humans they are subject to genetic defects resulting in the hereditary disorder galactosemia. When a defect is present, individuals are unable to metabolize the galactose derived from lactose (milk sugar) to glucose metabolites, often with resultant cataract formation, growth failure, mental retardation, or eventual death from liver damage. The genetic disturbance is expressed as a cellular deficiency of either galactokinase, causing a relatively mild disorder characterized by early cataract formation, or of galactose 1-phosphate uridylyl-transferase, resulting in severe disease.

Galactose is reduced to galactitol in a reaction similar to the reduction of glucose to sorbitol. Galactitol is the initiator of cataract formation in the lens and may play a role in the central nervous system damage. Accumulation of galactose 1-phosphate is responsible for liver failure; the toxic effects of galactose metabolites disappear when galactose is removed from the diet.

Oxidation and reduction interconversions result in formation of several additional sugars. **Glucuronic acid** is formed by oxidation of UDP-glucose catalyzed by **UDP-glucose dehydrogenase** and most likely follows. In humans glucuronic acid is converted to L-xylulose, the ketopentose excreted in **essential pentosuria,** and participates in detoxification by production of glucuronide conjugates.

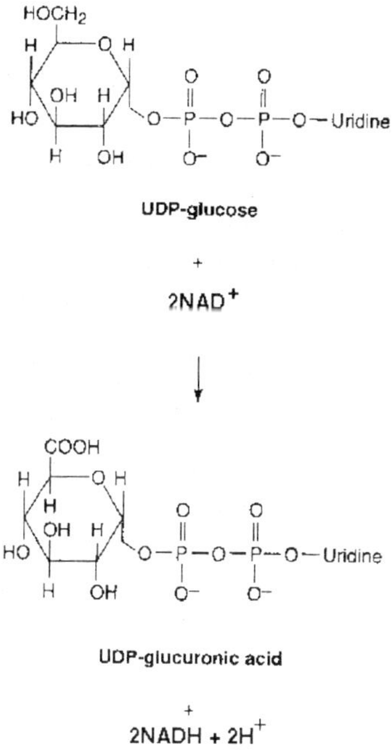

UDP-glucose

+

$2NAD^+$

↓

UDP-glucuronic acid

+

$2NADH + 2H^+$

Formation of UDP-glucuronic acid from UDP-glucose.

Glucuronic acid is a precursor of **L-ascorbic acid** in those animals that synthesize vitamin C. Free glucuronic acid can be metabolized by reduction with NADPH to L-gulonic acid, which is then converted by a two-step process through L-gulonolactone to L-ascorbic acid (vitamin C) in plants and most higher animals. Humans, other primates, and the guinea pig lack the enzyme that converts L-gulonolactone to L-ascorbic acid and therefore must satisfy their needs for ascorbic acid by its ingestion. Gulonic acid can also be oxidized to 3-ketogulonic acid and decarboxylated to L-xylulose. L-Xylulose is reduced to xylitol, reoxidized to D-xylulose, and phosphorylated with ATP and an appropriate kinase to xylulose 5-phosphate. The latter compound can then reenter the pentose phosphate pathway described previously. The glucuronic acid pathway represents another pathway for oxidation of glucose. This pathway operates in adipose tissue, and its activity can be increased in tissue from starved or diabetic animals.

7.9 Glycoproteins

Glycoproteins have been restrictively defined as conjugated proteins that contain, as a prosthetic group, one or more saccharides lacking serial repeat unit and bound covalently to a peptide chain.

The functions of glycoproteins in the human are of great interest. Glycoproteins in cell membrances may have an important role in the group behaviour of cells and other important biological functions of the membrane. Glycoproteins form a major part of the mucus that is secreted by epithelial cells, where they perform an important role in lubrication and in the protection of tissues lining the body's ducts. Many other proteins secreted from cells into extracellular fludis are glycoproteins. These proteins include hormones found in blood, such as follicle-stimulating hormone, luteinizing hormone, and chorionic gonadotropin; and plasma proteins such as the orosomucoids, ceruloplasmin, plasminogen, prothrombin, and immunoglobulins.

The percentage of carbohydrate in glycoproteins is highly variable. Some glycoproteins such as IgG contain low amounts (4%) of carbohydrate by weight, while glycophorin, the human red cell membrane glycoprotein, contains 60% carbohydrate. Human ovarian cyst glycoprotein contains 70% carbohydrate, and human gastric glycoprotein is 82% carbohydrate. The carbohydrate can be distributed fairly evenly along the polypeptide chain or concentrated in defined regions. For example, in human glycophorin. A the carbohydrate is found in the NH_2-terminal half of the polypeptide chain that lies on the outside of the cellular membrane.

The carbohydrate attached at one or at multiple points along a polypeptide chain usually contains less than 12-15 sugar residues. In some cases the carbohydrate component consists of a single sugar moiety, as in the submaxillary gland glycoprotein (single *N*-acetyl-α-D-galactosaminyl residue) and in some types of mammalian collagens (single α-D-galactosyl residue). In general, glycoproteins contain sugar residues in the D form, except for L-fucose, L-arabinose, and L-iduronic acid. A glycoprotein from different animal species often has an identical primary structure in the protein component, but a variable carbohydrate component. This heterogeneity of a given protein may even be true within a single organism. For example, pancreatic ribonuclease A and B forms have identical amino acid sequences and a similar kinetic specificity toward substrates, but they differ significantly in their carbohydrate composition.

Certain generalities about the structure of glycoproteins have emerged. Covalent linkage of sugars to the peptide chain is a central part of glycoprotein structure, and only a limited number of bonds are found. The three major types of **glycopeptide bonds,** are *N*-glycosyl to **asparagine** (Asn.), *O*-glycosyl to **serine** (Ser) or **threonine** (Thr), and *O*-glycosyl to **5-hydroxylysine.** The latter linkage, representing the carbohydrate side chains of either a single galactose or the disaccharide glucosylgalactose covalently bonded to hydroxylysine, is generally confined to the collagens. The other two linkages occur in a wide variety of glycoproteins. Of the three major types, only the *O*-glycoside linkage to serine or threonine is labile to alkali cleavage. By this procedure two types of oligosaccharides (simple and complex) are released. Examination of the simple class from porcine submaxillary mucins reveals some general structural feaures. A core structure exists, consisting of galactose (Gal) linked $\beta(1\to3)$ to *N*-acetylgalactosamine (GalNAc) *O*-glycosidically linked to serine or threonine residues. Residues of L-fucose (Fuc), sialic acid (NeuAc), and another *N*-acetylgalactosamine are found at the nonreducing periphery of this class of glycopeptides. The general structure of this type of glycopeptide is as follows:

$$\begin{array}{ccccc} \text{GalNAc} \xrightarrow{1,3} & \text{Gal} & \xrightarrow{1,3} & \text{GalNAc} & \to O-\text{Ser/Thr} \\ & \uparrow 1,2 & & \uparrow 2.6 & \\ & \text{Fuc} & & \text{NeuAc} & \end{array}$$

More complex heterosaccharides are also linked to peptides via serine or threonine residues and are exemplifed by the blood group substances. Study of these determinants has shown how complex and variable these structures are, as well as how the oligosaccharides of cell surfaces are assembled and how that assembly pattern is genetically determined. Certain common structural features of the oligosaccharide *N*-glycosidically linked to asparagine have also emerged. These glycoproteins commonly contain a core structure consisting of mannose (Man) residues linked to *N*-acetylglucosamine (GlcNAc) in the following structure:

$$(\text{Man})_n \xrightarrow[1,4]{} \text{Man} \xrightarrow[1,4]{} \text{GlcNAc} \xrightarrow[1,4]{} \text{GlcNAc} \rightarrow \text{Asn}$$

7.10 Formation of Glycosaminoglycan

The polysaccharide chains of proteoglycans are assembled by sequential action of a series of glycosyltransferases in the endoplasmic reticulum, which catalyze the transfer of a monosaccharide from a nucleotide sugar to an appropriate acceptor, either the nonreducing end of another sugar or a polypeptide. Since the biosynthesis of the chondroitin sulfates is most thoroughly understood, this pathway will be discussed as the prototype for glycosaminoglycan formation.

Formation of the core protein of the chondroitin sulfate proteoglycan is the first step in this process, followed by assembly of the polysaccharide chains catalyzed by six different glycosyltransferases in the lumen of the endoplasmic reticulum. Strict substrate specificity is required for completion of the unique tetrasaccharide linkage region. Polymerization then results from the concerted action of two glycosyltransferases, an *N*-acetylgalactosaminyltransferase and a glucuronosyltransferase, which alternately add the two monosaccharides, forming the characteristic repeating disaccharide units. Sulfation of *N*-acetylgalactosamine residues in either the 4 or 6 position apparently occurs along with chain elongation. The sulfate donor, as in other biological systems, is 3′-phosphoadenosine 5′-phosphosulfate (PAPS), which is formed from ATP and sulfate in two steps.

Synthesis of other glycosaminoglycans requires additional transferases specific for the sugar and linkages found in these molecules. Completion often involves modifications in addition to *O*-sulfation, including epimerization, acetylation, and *N*-sulfation. Interestingly, the epimerization of D-glucuronic acid to L-iduronic acid occurs after incorporation into the polymer chain and is coupled with the process of sulfation.

Synthesis of both proteoglycans and glycoproteins is regulated by the same mechanism at the level of hexosamine synthesis. The fructose 6-phosphate-glutamine transamidase reaction is subject to feedback inhibition by UDP-*N*-acetylglucosamine, which is in equilibrium with UDP-*N*-acetylgalactosamine. More specific to proteoglycan synthesis, the levels of UDP-xylose and UDP-glucuronic acid are stringently controlled by the inhibition by UDP-xylose of the UDP-glucose dehydrogenase conversion of UDP-glucose to UDP-glucuronic acid. Since xylose is the first sugar added during synthesis of chondroitin sulfate, dermatan sulfate, heparin, and heparan sulfate, the earliest effect of decreased core protein synthesis would be accumulation of UDP-xylose, which aids in maintaining a balance between synthesis of protein and polysaccharide moieties of these complex macromolecules.

Proteoglycans, like glycoproteins, are degraded by the sequential action of proteases and **glycosidases,** as well as **deacetylases** and **sulfatases.** Much of the information about metabolism and degradation of proteoglycans has been derived from the study of **mucopolysaccharidoses**. This group of human genetic disorders is characterized by accumulation in tissues and excretion in urine of oligosaccharide products derived from incomplete breakdown of the proteoglycans, due to a deficiency of one or more lysosomal hydrolases. In the diseases for which the biochemical defect has been identified, a product accumulates that has a nonreducing terminus that would have been the substrate for the deficient enzyme.

Although proteoglycans continue to be defined on the basis of the glycosaminoglycan chain they contain, new ones are increasingly being described based largely on functional properties or location. Aggrecan and versican are the predominant extracellular species; syndecan, CD44, and thrombomodulin are integral membrane proteins; neurocan, brevican, cerebrocan, and phosphacan are largely restricted to the nervous system; while many proteo-glycans (i.e., aggrecan, syndecan, and betaglycan) carry two types of glycosaminoglycan chains, whose size and ratio may change with development age, or disease. Thus it appears that the versatile structure of these abundant carbohydrate-containing molecules is well exploited by cells in many as yet undiscovered ways.

8

Amino Acid Metabolism

8.1 Introduction

Amino acids are the monomeric building blocks of proteins. **Glutamate and glutamine** are also the only source of nitrogen atom available to mammals for the biosynthesis of other α-amino acids and all other biologically important non-protein nitrogen-containing compounds, e.g. amines, haem, purine and pyrimidine bases.

Excess dietary amino acids, unlike glucose or fatty acids, are not stored. Amino acids which are surfeit to the demands of synthetic pathways are not excreted but are utilized as sources of energy, either directly or indirectly. About 10-15% of human energy requirements are derived from the oxidation of amino acids. It is, however, only the chains of carbon atoms, called **carbon skeletons,** which are useful as metabolic fuels and not the α-amino groups. **Deamination** of some amino acids results in the production of toxic ammonia. Ammonia is converted to **urea,** a non-toxic product at physiological levels, which provides a convenient soluble vehicle for the disposal of amino nitrogen atoms.

In man, ~50% of body protein is replaced every 5–6 months. This turnover involves a controlled balance between degradation and synthesis. The liver is the major site for both the biosynthesis and breakdown of amino acids in man. Since there are 20 amino acids found in mammalian proteins, there are about 20 pathway branches concerned with their catabolism and a similar number involved in their anabolism. For this reason, only the underpinning principles of amino acid metabolism are considered here.

A cell in general will utilize thousands of different proteins which constitute over half of the total dry mass of the cell. Proteins are central to the cell's well-being, growth and development.

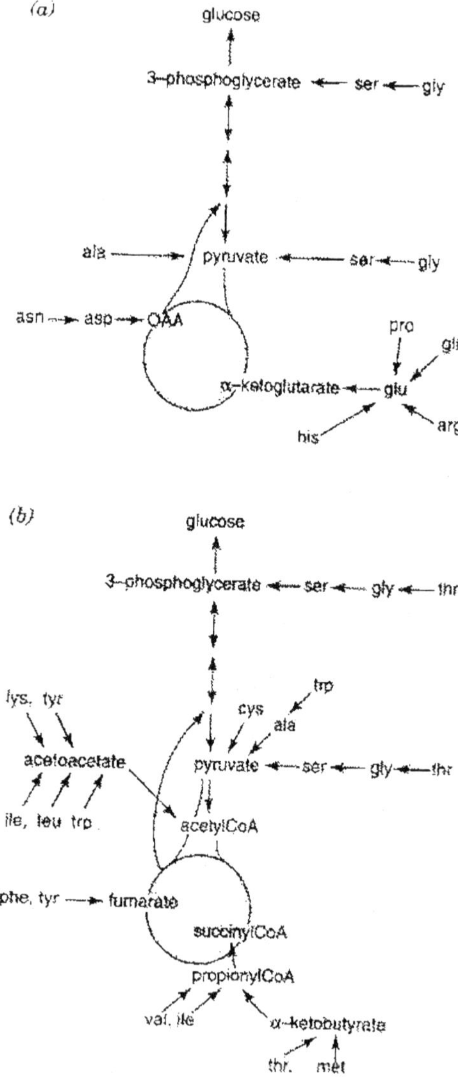

Metabolic fate of (*a*) nonessential amino acids; (*b*) essential amino acids plus cysteine and tyrosine.

Proteins are the final recipients of the information contained within the exons of the genome. RNA is transcribed from the gene and in mammalian cells is processed into mRNA which directs the assembly of amino acids into the primary structure of specific proteins. mRNA is translated by the ribosomes from which a polypeptide chain emerges.

This chapter considers the degradation of amino acids, disposal of amino nitrogen as urea and the *de novo* synthesis of amino acids for protein synthesis. The chapter also discusses the employment of amino acids in the translation of mRNA into a polypeptide chain.

Molecular nitrogen, N_2, exists in the atmosphere in great abundance. Before it can be utilized by animals it must be "fixed," that is, reduced from N_2 to NH_3 by microorganisms, plants, and electrical discharge from lightning. Ammonia is then incorporated into amino acids and proteins, and these become part of the food chain. Humans can synthesize only 11 of the 20 amino acids needed for protein synthesis. Those that cannot be synthesized *de novo* are termed "**essential**" because they must be obtained from dietary foodstuffs that contain them.

Amino acids are the sources of all nitrogen and some carbon atoms of **heterocyclic compounds**. The atoms of the **pyrimidine ring** are derived from two amino acids, aspartate and glutamine, and CO_2. Atoms N-1, C-4, C-5 and C-6 come from aspartate, C-2 from CO_2 and N-3 from glutamine.

The atoms of the **purine ring** are derived from three amino acids, aspartate, glutamine and glycine, CO_2 and formly groups. Atom N-1 comes from aspartate, N-3 and N-9 from glutamine, C-4, C-5 and N-7 are the framework atoms of glycine and C-6 from CO_2. Atoms C-2 and C-8 are donated by two **formylation reactions** in which different **folate dervatives,** N^{10}-formyltetrahydrofolate and N^5, N^{10}-methenyltetrahydrofolate respectively, are formyl group donors.

Biosynthesis does not produce free heterocycline compounds but nucleotides. Although purines contain a pyrimidine ring, the synthetic pathways are distinct with the imidazole ring being initially constructed. Since synthesis of pyrimidine and purine nucleotides *de novo* (anew) is energetically expensive, heterocyclic bases from dietary sources or recycled from the turnover of nucleic acids by the **salvage pathways,** involving **hypoxanthine-guanine phosphoribosyltransferase,** facilitate considerable saving in ATP.

8.2 Transfer of Amino Groups

Most amino acids used by the body to synthesize protein or as precursors for amino acid derivatives are obtained from the diet or from protein turnover. When necessary, nonessential amino acids are synthesized from α-keto acid precursors via transfer of a preexisting amino group from another amino acid by **aminotransferases,** also called **transaminases**. Transfer of amino groups also occurs during degradation of amino acids. How the amino group of alanine is transferred to α-ketoglutarate to form glutamate. In this reaction the pyruvate produced provides carbons for gluconeogenesis or for energy production via the TCA cycle. This reaction is necessary since ammonia cannot enter the urea cycle directly from alanine but can be donated by glutamate. The opposite reaction would occur if there were a need for alanine for protein synthesis that was not being met by dietary intake or protein turnover. Transmination involving essential amino acids to normally unidirectional since the body cannot synthesize the equivalent α-keto acid. The resulting α-ketoisovalerate is further metabolized to succinyl CoA. Transamination is the most common reaction involving free amino acids, and only threonine and lysine do not participate in an aminotransferase reaction. An obligate amino and α-keto acid pair in all of these reactions is glutamate and α-ketoglutarate. This means that amino group transfer between alanine and aspartate would have to occur via coupled reactions, with a glutamate intermediate. The equilibrium constant for aminotransferases is close to one so that the reactions are freely reversible. When nitrogen excretion is impaired and **hyperammonemia** occurs, as in liver failure, amino acids, including the essential amino acids, can be replaced in the diet by α-keto acid analogs, with the exception of threonine and lysine as mentioned above. The α-keto acids are transmitted by aminotransferases to produce the different amino acids.

Tissue distribution of some of the aminotransferase family is used diagnostically by measuring the release of a specific enzyme during tissue damage; for instance, the presence of glutamate oxaloacetate aminotransferase in plasma is a sign of liver damage.

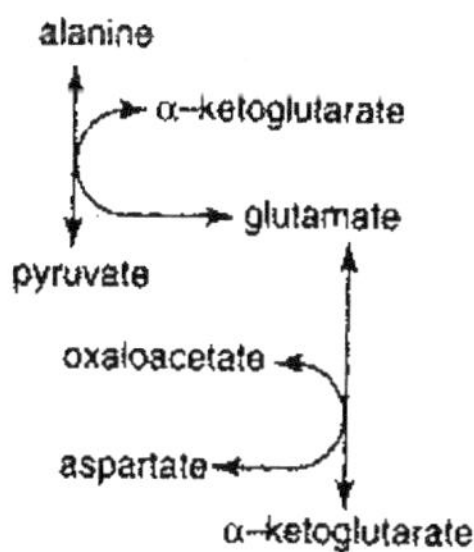

A coupled transmination reaction.

Transfer of amino groups occurs via enzyme-associated intermediates derived from **pyridoxal phosphate,** the functional form of vitamin B_6. The active site of the "resting" aminotransferase contains pyridoxal phosphate covalently attached to an ε-amino group of a lysine residue that forms part of the amino acid chain of the transferase. The complex is further stabilized by ionic and hydrophobic interaction. The linkage, –CH=N–, is called a **Schiff base.** The carbon originates in the aldehyde group of pyridoxal phosphate, and the nitrogen is donated by the lysine residue. When a substrate amino acid, ready to be metabolized, approaches the active site, its amino group displaces the lysine ε-amino group and a Schiff base linkage is formed with the amino group of the amino acid substrate. At this point the pyridoxal phosphate-derived molecule is no longer covalently attached to the enzyme but is held in the active site only by ionic and hydrophobic interactions between it and the protein. The Schiff base linkage involving the amino acid substrate is in tautomeric equilibrium between an aldimine, $-CH=N-CHR_2$, and a ketimine, $-CH_2-N=CR$. Hydrolysis of the ketimine liberates an α-keto acid, leaving the amino group as part of the pyridoxamine structure. A reversal of the process is now possible; an α-keto acid reacts with the amine group, the double bond is shifted, and then hydrolysis liberates an amino acid. Pyridoxal phosphate now reforms its Schiff base with the "resting" enzyme. Most pyridoxal phosphate-requiring reactions involve transamination, but the ability of the Schiff base to transfer electrons between different atoms allows this cofactor to participate.

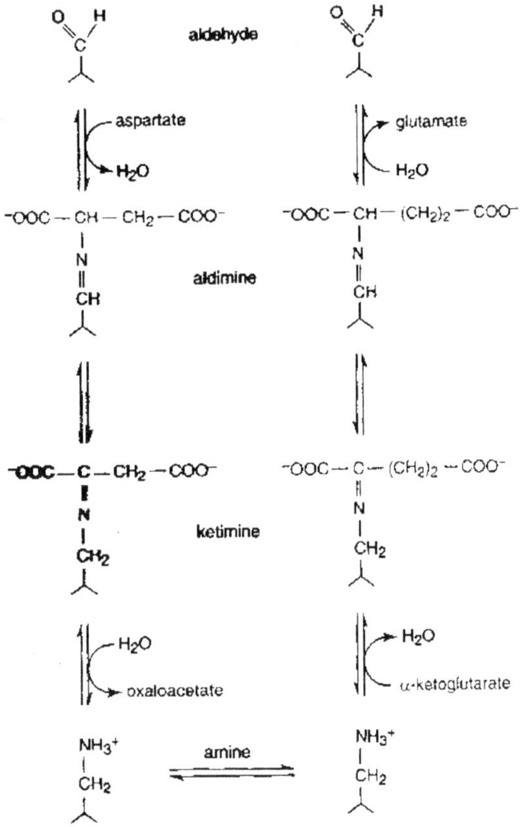

Different forms of pyridoxal phosphate during a transamination reaction.

8.3 Interorgan Nitrogen Transport

Whole cells die on a regular and planned basis, and their component molecules are metabolized. This "**planned cell death**" is called **apoptosis.** Individual proteins also undergo regular turnover under normal conditions. Even though the reactions involved in intracellular protein degradation have been identified, an understanding of the regulation of protein degradation is in its infancy. The half-life of a protein can be an hour or less, such as for ornithine decarboxylase, phosphokinase C, and insulin, several months for hemoglobin and histones, or the life of the organism for the crystallins of the lens. The majority, however, turn over every few days. Selection of a particular protein molecule for degradation is not well understood but may, in many cases, occur by "marking" with covalently bound molecules of an oligopeptide, termed **ubiquitin.** Ubiquitin contains 76 amino acid residues and is attached via its C-terminal glycine residue to the terminal amino group and to lysine residues in the protein to be marked for degradation. This is a nonlysosomal, ATP-dependent process and requires a complex of three enzymes known as ubiquitin protein ligase. Recently, ubiquitination and protein degradation have been found to regulate the cell cycle by influencing the availability of proteins required in the S and G_1 phases. Other protein degradation occurs in the lysosomes, or extralysosomally by calciumdependent enzymes.

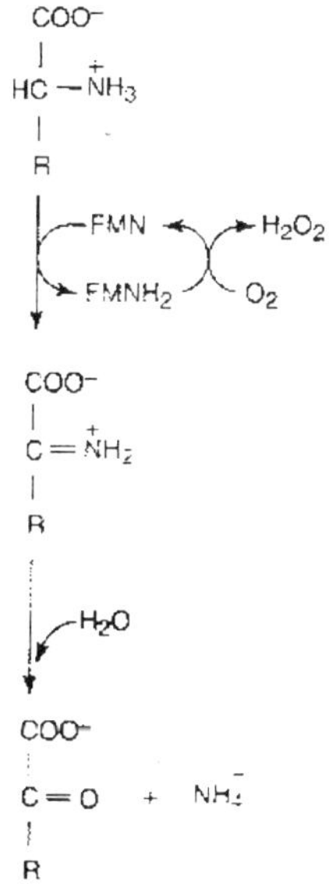

The majority of protein, and consequently of amino acids, is in skeletal muscle. Under conditions of energy need, this protein is degraded and amino groups from the amino acids are transferred to glutamine and alanine and transported to liver or kidney. Urea is produced in liver and ammonia (from glutamine) in kidney. Carbon skeletons are either used for energy or transported to the liver for gluco-neogenesis. Muscle protein responds to conditions such as starvation, trauma, burns, and septicemia, by undergoing massive degradation. Of the amino acids released, most important as a source of fuel are **branched-chain amino acids** (valine, leucine, and isoleucine). The first step in

COO^- | $HC-\overset{+}{N}H_3$ | R

FMN → H_2O_2 ; $FMNH_2$ ← O_2

COO^- | $C=\overset{+}{N}H_2$ | R

H_2O

COO^- | $C=O$ + NH_4^+ | R

Reaction of L-amino acid oxidase, a flavoprotein.

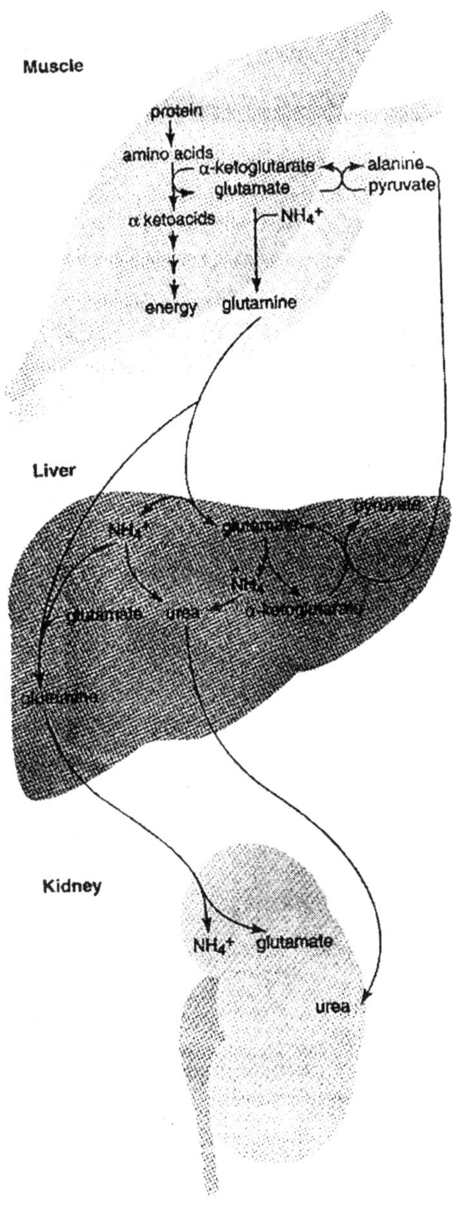

Major pathways of interorgan nitrogen transport following muscle proteolysis.

their degradation is transamination, which occurs almost exclusively in muscle. Protein is, of course, degraded throughout the body, but muscle is by far the greatest source of free amino acids for metabolism.

The main destination of glutamine and alanine in the blood is the liver. Here ammonia is released by alanine aminotransferase, glutaminase, and glutamate dehydrogenase. Glutamate dehydrogenase not only releases ammonia but also produces NADH and α-ketoglutarate, a glucogenic intermediate. Under conditions of energy need these products are very beneficial. Many tumors produce a condition called **cachexia,** characterized by wasting of muscle. This is caused not at the level of regulation of the rate of muscle protien breakdown, but rather by an increase in the rate at which liver removes amino acid from plasma, which, in turn, has a potentiating effect on muscle proteolysis. When circulating glucagon concentration is high (a signal that carbon is required by the liver for gluconeogenesis), it also potentiates amino acid metabolism by stimulating amino acid uptake by the liver.

Some glutamine and alanine is taken up by the kidney. Ammonia is released by the same enzymes that are active in liver, protonated to ammonium ion and excreted. When acidosis occurs the body shunts glutamine from liver to kidney to **conserve bicarbonate,** since formation of urea, the major mechanism for removal of NH_4^+, requires bicarbonate. To avoid use and excretion of this anion as urea during acidosis, uptake of glutamine by liver is suppressed, and more is transported to kidney for excretion as ammonium ion.

8.4 Tyrosine and Phenylalanine

Tyrosine and **phenylalanine** are discussed together, since tyrosine results from hydroxylation of phenylalanine and is the first product in phenylalanine degradation. Because of this, tyrosine is not usually considered to be essential, whereas phenylalanine is. Three-quarters of ingested phenylalanine is metabolized to tyrosine. This is catalyzed by **phenylalanine hydroxylase**, which is **tetrahydrobiopterin dependent**. This reaction occurs only in the

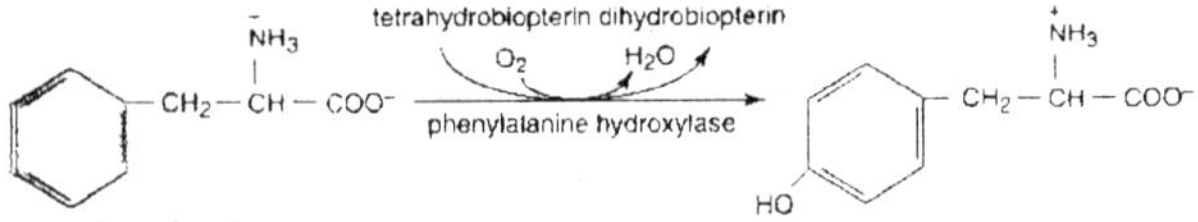

Phenylalanine **Tyrosine**

Phenylalanine hydroxylase.

Phenylalanine hydroxylase.

direction of tyrosin formation, and phenylalanine cannot be synthesized from tyrosine. **Biopterin,** unlike folic acid which it resembles in containing pteridine ring is not a vitamin. It is synthesized from GTP.

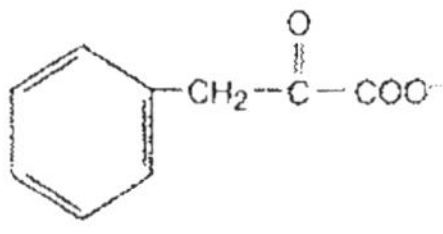

Phenylpyruvate

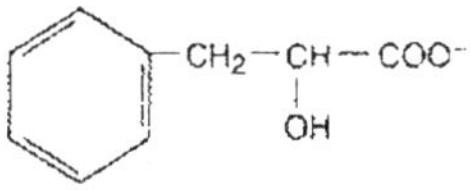

Phenyllactate

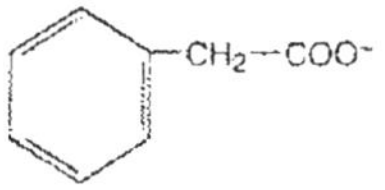

Phenylacetate

Minor products of phenylalanine metabolism.

The first step in metabolism of tyrosine is transamination by **tyrosine aminotransferase** to *p*-hydroxyphenylpyruvate. The enzyme is inducible, its synthesis being increased by glucocorticoids and dietary tyrosine. ***p*-Hydroxyphenylpyruvate oxidase** produces **homogentistic acid.** This complex reaction involves decarboxylation, oxidation, migration of the carbon side chain, and hydroxylation. Ascorbic acid is required for at least one of these activities, but all four are catalyzed by the one enzyme. The aromatic ring is next cleaved by an iron-containing enzyme, homogentisate oxidase, to maleylacetoacetate.

Now let us study about the disorders of Tyrosine metabolism.

Tyrosinemias

The absence or deficiency of tyrosine aminotransferase produces accumulation and excretion of tyrosine and metabolites. The disease, oculocutaneous or type II tyrosinemia, results in eye and skin lesions and mental retardation. Type I, hepatorenal tyrosinemia, is more serious, involving liver failure, renal tubular dysfunction, rickets, and polyneuropathy, caused by a deficiency of fumarylacetoacetate hydrolase. Accumulation of fumaryl-acetoacetate and maleylacetate, both of which are alkylating agents, can lead to DNA alkylation and tumorigenesis. Both diseases are autosomal recessive and rare.

Alcaptonuria

The first condition identified as an "inborn error of metabolism" was alcaptonuria. Individuals deficient in homogentisate oxidase excrete almost all ingested tyrosine as the colorless homogentistic acid in their urine. This auto-oxidizes to the corresponding quinone, which polymerizes to form an intensely dark color. Concern about the dark urine is the only consequence of this condition early in life. Homogentisate is slowly oxidized to pigments that are deposited in bones, connective tissue, and other organs, a condition called ochronosis because of the ochre color of the deposits. This is thought to be responsible for the associated arthritis, especially in males. The study of alcaptonuria by Archibald Garrod, who first indicated its autosomal recessive genetic basis, includes an unusual historical description of the iatrogenic suffering of the first patient treated for the condition, which is frequently benign.

Albinism

Skin and hair color are controlled by an unknown number of genetic loci in humans and exist in infinite variation; in mice, 147 genes have been identified in color determination. Many conditions have been described in which the skin has little or no pigment. The chemical basis is not established for any except classical albinism, which results from a lack of tyrosinase. Lack of pigment in the skin makes albinos sensitive to sunlight, increasing carcinoma of the skin in addition to burns; lack of pigment in the eyes may contribute to photophobia.

8.5 Tryptophan Metabolism

Metabolism of **tryptophan** has many branch points. The dominant or oxidative pathway of tryptophan in the human starts with oxidation of tryptophan to *N*-formylkynurenine by a heme-containing enzyme, **tryptophan dioxygenase,** also called **tryptophan pyrrolase** or **tryptophan oxygenase,** because the pyrrole ring is cleaved in the reaction. Tryptophan dioxygenase is induced by glucocorticoids and glucagon. It is found in liver; other tissues contain a similar enzyme called indolamine dioxygenase, which is less substrate specific. Formamidase then hydrolyzes formylkynurenine to formate and kynurenine. At this point the pathway begins to branch. In the dominant pathway, reactions lead to 3-hydroxykynurenine, 3-hydroxyanthranilic acid and alanine, amino-carboxymuconic semialdehyde, and by decarboxylation, to aminomuconic semialdehyde. This can be further metabolized in several steps to glutarate and eventually acetoacetyl CoA, or recyclized nonenzymatically to **picolinic acid,** which is excreted in the urine.

Tryptophan is the precursor of approximatly 50% of the body's pyridine nucleotides. The rest is obtained from the diet. The branch point leading to nictolinate mononucleotide at the stage of amino carboxymuconic semialdehyde. The enzyme that forms 2-aminomuconic semialdehyde, picolinate carboxylase from this compound has a low K_m and is easily saturated with substrate.

Since picolinate carboxylase has low activity in liver, some amino-carboxymuconic semialdehyde is cyclized in a nonenzymatic reaction to quinolinic acid. Phosphoribosylpyrophosphate provides a ribonucleotide moiety and the final step is a decarboxylation leading to nicotinate mononucleotide. Note that the nicotinic acid ring is synthesized as a part of a nucleotide. Because **kynurenine hydroxylase** is inhibited by estrogen, women are more susceptible to **pellagra**, the disease produced by niacin deficiency (from the Italian *pelle,* skin, and *agra,* rough).

Many enzymes in this lengthy pathway are pyridoxal phosphate dependent. **Kynureninase** is one of them and is affected by a vitamin B_6 deficiency, resulting in excess kynurenine and xanthurenate excretion and giving urine a greenish-yellow colour. This is a diagnostic symptom of vitamin B_6 deficiency.

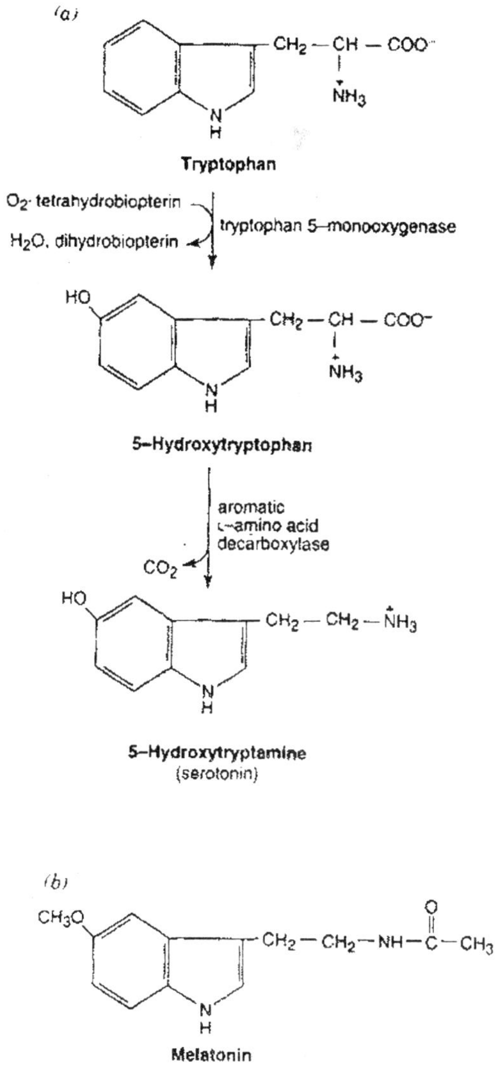
(a)
CH2 — CH — COO⁻
NH3
N
H
Tryptophan
O2· tetrahydrobiopterin
H2O, dihydrobiopterin
tryptophan 5–monooxygenase
HO
CH2 — CH — COO⁻
NH3
5–Hydroxytryptophan
aromatic
L–amino acid
decarboxylase
CO2
HO
CH2 — CH2 — NH3
5–Hydroxytryptamine
(serotonin)
(b)
CH3O
O
CH2 — CH2 — NH — C — CH3
Melatonin

(*a*) Synthesis of serotonin (5-hydroxytryptamine) and (*b*) structure of melatonin.

Another pathway that kynurenine can follow is transamination and condensation of the side chain to form a two-ring compound, kynurenic acid. **Kynurenic acid,** its decarboxylated metabolite **kynuramine,** and **quinolinate** have all been shown to act as tryptophan-derived neurotransmitters, possibly as antiexcitotoxics and anticonvulsives.

Serotonin (5-hydroxytryptamine) results from hydroxylation of tryptophan by a tetrahydrobiopterin-dependent enzyme and decarboxylation by a pyridoxal phosphate-containing enzyme. It is a neurotransmitter in brain and causes contraction of smooth muscle of arterioles and bronchioles. It is found widely in the body and may have other physiological roles. **Melatonin,** a sleep-inducing molecule, is ***N*-acetyl-5-methoxytryptamine**. The acetyltransferase needed for its synthesis is present in pineal gland and retina. Melatonin is involved in regulation of circadian rhythm, being synthesized mostly at night. It appears to function by inhibiting synthesis and secretion of other neurotransmitters such as dopamine and GABA.

Ingestion of foods rich in tryptophan leads to sleepiness because serotonin is also sleep-inducing. Reducing availability of tryptophan in the brain can interfere with sleep. Tryptophan availability is reduced when other amino acids compete with it for transport through the blood-brain barrier. Elevated plasma concentrations of other amino acids, after a high-protein meal, diminish transport of tryptophan and induce wakefulness. The sleep-inducing effect of carbohydrates is due to decreased plasma amino acid levels, since carbohydrate stimulates release of insulin, and insulin causes removal of amino acids from plasma and uptake into muscle. This alleviates competition and increases the amount of tryptophan that can enter the brain. Strangely, extra serotonin appears to lead to sleepiness in females, but only calmness in males.

8.6 Glutathione

Glutathione, the tripeptide γ-glutamylcysteinylglycine, has several important functions. It is a reductant, conjugated to drugs to make them more water soluble involved in transport of amino acids across cell membranes, part of some leukotriene structures, a cofactor for some enzymatic reactions, and an aid in the rearrangement of protein disulfide bonds.

Glutathione as reductant is very important in maintaining stability of erythrocyte membranes. Its sulfhydryl group can be used to reduce peroxides formed during oxygen transport. The resulting oxidized form of GSH consists of two molecules joined by a disulfide bond. This is reduced to two molecules of GSH at the expense of NADPH. The usual steadystate radio of GSH to GSSG in erythrocytes is 100:1.

Conjugation of drugs by glutathione, often after a preliminary reaction catalyzed by cytochrome P450, renders them more polar for excretion.

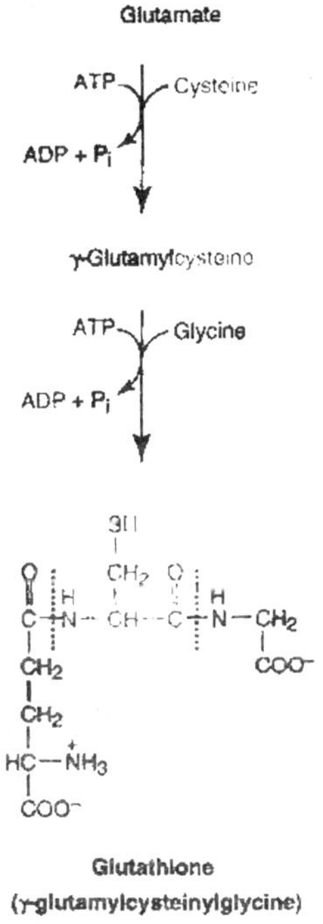

Glutamate

ATP, Cysteine → ADP + P_i

γ-Glutamylcysteine

ATP, Glycine → ADP + P_i

Glutathione
(γ-glutamylcysteinylglycine)

Synthesis of glutathione.

Glutathione is synthesized by formation of the dipeptide γ-glutamylcysteine and the subsequent addition of glycine. Both reactions require activation of carboxyl groups by ATP. Synthesis of glutathione is largely regulated by cysteine availability.

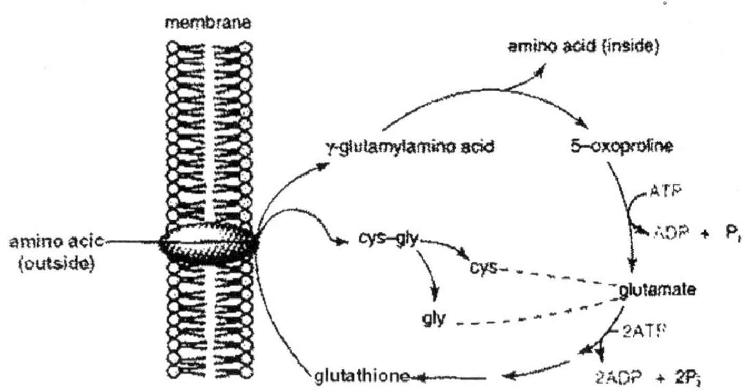

γ-Glutamyl cycle for transporting amino acids.

There are several mechanisms for transport of amino acids across cell membranes. Many are symport or antiport mechanisms and are coupled to sodium transport. The γ-**glutamyl cycle** is an example of "group transfer" transport. It is more energy-requiring than other mechanisms, but is rapid and has high capacity, and functions in the kidney and some other tissues. It is particularly important in renal epithelial cells.

The enzyme γ-**glutamyl transpeptidase** is located in the cell membrane. It shuttles GSH to the cell surface to interact with an amino acid. γ-Glutamyl amino acid is transported into the cell, and the complex is hydrolyzed to liberate the amino acid. Glutamate is released as 5-**oxoproline,** and cysteinylglycine is cleaved to its component amino acids. To regenerate GSH glutamate is reformed from oxoproline in an ATP-requiring reaction, and GSH is resynthesized from its three component parts. Three ATPs are used in the regeneration of glutathione, one in formation of glutamate from oxoproline and two in formation of the peptide bonds.

When the body encounters toxic conditions such as peroxide formation, ionizing radiation, alkylating agents, or other reactive intermediates, it is beneficial to increase the level of GSH. Cysteine and methionine have been administered as GSH precursors, but they have the disadvantage of being precursors of an energy-expensive pathway to GSH. A more promising approach is administration of a soluble diester of GSH, such as γ-(α-ethyl) glutamy-lcysteinyl-ethylglycinate.

Very premature infants have a very low concentration of cysteine because of low cystathionase activity in liver. This keeps the GSH concentration low and makes them more susceptible to oxidative damage, especially from hydroperoxides formed in the eye after hyperbaric oxygen treatment. Under certain circumstances, such as rendering tumor cells more sensitive to radiation or parasites more sensitive to drugs, it is desirable to lower GSH levels. This can be achieved by the glutamate analog **buthionine** sulfoximine as a competitive inhibitor of GSH synthesis.

9

DNA: Structure and Function

9.1 Introduction

DNA is the macromolecule that ultimately controls every aspect of cellular function, primarily through protein synthesis. DNA exercises this control as suggested by the sequence.

$$\circlearrowleft \text{DNA} \longrightarrow \text{RNA} \longrightarrow \text{protein}$$

Flow of biological information is clearly from one class of nucleic acid to another, from DNA to RNA, with only minor exceptions, and from there to protein. For this transfer of information to occur faithfully, each preceding macromolecule serves as a structure-specifying template for the synthesis of the subsequent member in the sequence.

In addition to regulating cellular expression, DNA plays an exclusive role in heredity. This role is suggested by a circular arrow engulfing DNA, which depicts DNA as a replicon, a molecule that can undergo self-replication. Replication permits DNA to make copies of itself as a cell divides and bestows them to daughter cells, which can thus inherit every property and characteristics of the original cell. Thus DNA ultimately determines the properties of a living cell by regulating expression of biological information, primarily by control of protein synthesis, and transfers biological information from one generation to the next; that is, it transmits genetic information.

One of the striking aspects of natural order is the sense of unity that exists between members of successive generations in each species. An almost totally stable bank of information must always be preserved and passed from one generation to the next if individual species are to maintain their identities relatively unchanged over millions of years. It is well established that this bank of genetic information takes the form of a macromolecule, deoxyribonucleic acid (DNA), which serves as the carrier of genetic information in both prokaryotes and eukaryotes. DNA exhibits a rare purity of function by being one of the few macromolecules known to perform, with only minor exceptions, the same basic functions across species barriers.

Properties of cells are to a large extent determined by their constituent proteins. Many proteins are indispensable structural components of cells. Other proteins, such as enzymes and certain hormones, are functional and determine many of the biochemical properties of the cell. As a result, factors that control which proteins a cell may synthesize, at what quantities, and with which sequence are the same factors that primarily determine function and destiny of every living cell.

The first convincing suggestion that DNA is the genetic material was made during the mid-1940s. The experiment involved transformation of a type of pneumococcus, surrounded by a polysaccharide capsule and referred to as S form because of its property of forming colonies with smooth-looking cellular perimeters, to a mutant without capsule, called R form, which forms colonies with rough-looking outlines. These two forms are genetically distinct and cannot interconvert spontaneously. **Transformation** experiments demonstrated that an extract of pure DNA from S form, when incorporated into R form of pneumococcus, conveyed to R form the specific property of synthesizing the characteristic polysaccharide capsule. Furthermore, bacteria transformed from R form to S form maintained the property of synthesizing capsule over succeeding generations. It was thus demonstrated that DNA was the **transforming agent,** as well as the material responsible for transmitting genetic information from one generation to the next. Almost three-quarters of a century elapsed from the time nucleic acid were discovered until their important biological role was generally recognized.

A striking characteristic of DNA is its ability to encode an enormous quantity of biological information. An undifferentiated mammalian fetal cell contains only a few picograms (10^{-12} g) of DNA. Yet this minute amount of material is sufficient to direct synthesis of as many as 100,000 distinct proteins that will determine the form and biochemical behaviour of a large variety of differentiated tissues in adult animals.

The compactness of information storage in DNA is unique. Even sophisticated memory elements of contemporary computers appear pitifully inadequate by comparison. How does DNA achieve such a supreme **coding effectiveness?** Answers must obviously be sought in the nature of its chemical structure. It turns out that this structure is not only consistent with the unique efficiency of DNA as a "memory bank" but also provides the basis for understanding how DNA eventually "translates" this information into proteins.

Traditional procedures of vaccination have used purified components of an infectious organism, dead or attenuated intact cells or viruses, to provide individuals with active immunity by eliciting production of specific antibodies. Many have been successful in providing protection against diseases such as polio, smallpox, whooping cough, typhoid fever, and diphtheria.

A prototype DNA vaccine has been developed. It consists of a naked DNA that encodes the nucleoprotein of the influenza virus. This gene is the same or very similar in many strains of this virus and should afford protection against all or most of them. Naked DNA, that is, DNA freed of all its naturally associated proteins, is used. It enters cells and can be expressed without need of a complex virus system. Results of its use in mice and nonhuman primates have been very encouraging. Naked-DNA vaccines appear to stimulate cell-mediated immunity and an antibody response.

9.2 Composition of DNA

The **base composition of DNA** varies considerably among species, particularly prokaryotes, which have a ragne of 25-75% in adenine-thymine content. This range narrows with evolution, reaching limiting values of about 45-53% in mammals.

DNA contains various **methylated bases.** These methylated derivates are present in all prokaryotic DNA molecules examined to date but are absent in certain eukaryotes such as yeast and insects. As a rule these bases are generated by action of methylases, Dam and Dem, following synthesis of DNA. Methyl groups are transferred from S-adenosylmethionine. **Dam methylase** selects adenine residues on GATC sequences for methylation.

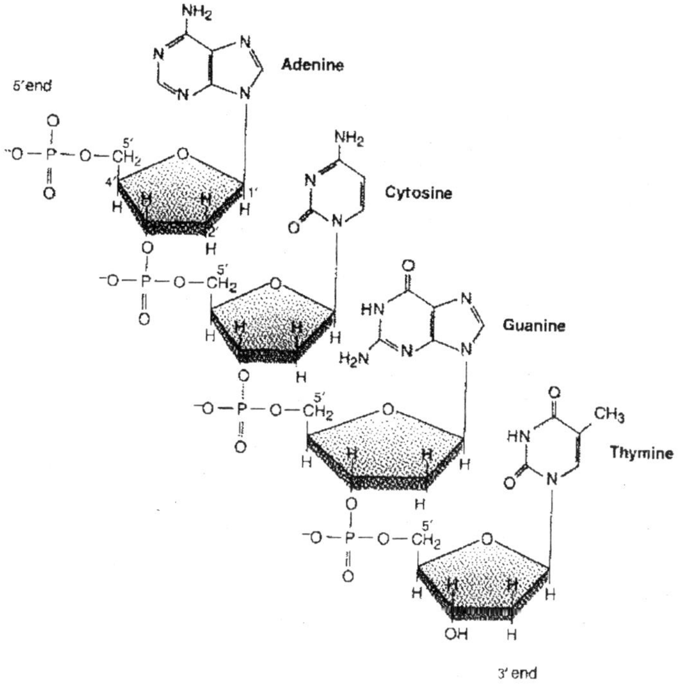

Structure of a DNA polynucleotide segment.

Polynucleotides are formed by joining of nucleotides by **phosphodiester bonds**. The phosphodiester bond is the formal analog of the peptide bond in proteins. It joins by esterification of two of the three OH groups of phosphoric acid, two adjoining nucleotide residues. Deoxyribose contains two free OH groups on the C-3′ and C-5′ atoms that can participate in formation of a phosphodiester bond. Indeed, the nucleotide residues in DNA are joined by **3′,5′-phosphodiester bonds.**

Many polynucleotides are linear polymers. The last nucleotide residues at opposite ends of the polynucleotide chain serve as the two terminals of the chain. It is apparent that these terminals are not structurally equivalent, since one of the nucleotides must terminate at a 3′-OH group and the other at a 5′-OH group. These ends of the polynucleotides are referred to as 3′ and 5′ termini, and they may be viewed as corresponding to the amino and carboxyl termini in proteins. Polynucleotides also exist as cyclic structures, which contain no free terminals.

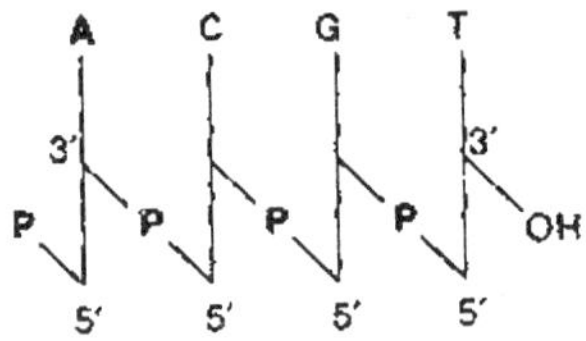

Shorthand form for structure of oligonucleotides.

Esterification between the 3′-OH terminus of a polynucleotide with its own 5′-phosphate terminus can produce a cyclic polynucleotide.

Long polymers of nucleotides joined by phosphodiester bonds are called polynucleotides. **Oligonucleotides** are shorter nucleotide-containing polymers. According to formal rules of nomenclature, however, polynucleotides are named by using roots derived from the names of corresponding nucleotides, and using the ending ylyl. Polynucleotide sequences are always read in the 5′ → 3′ direction, unless specified otherwise. For example, the polynucleotide segment in which the 5′ terminal is on the left of each nucleotide residue, should be named from left to right as

... deoxyadenylyl, deoxycytidylyl, deoxyguanylyl, deoxythymidylyl...

However, use of complete chemical names is cumbersome and abbreviations are generally preferred. For example, the oligonucleotide is usually referred to as dAdCdGdT, and a polynucleotide containing only one kind of nucleotide, for example, dA, may be written as poly(dA). Oligo- and polynucleotide structures are also written out in shorthand.

The specific sequence of bases along a polynucleotide chain determines its biological properties. Although the structure of the nucleic acid bases had been known for many years, the polymeric structure initially proposed for DNA was one of the classical errors in the history of biochemistry. Experimental data obtained from partially degraded samples of DNA, and several misconceptions, led to the erroneous conclusion that DNA consisted of repeating tetranucleotide units. Each tetranucleotide supposedly contained equimolar quantities of the four common bases. These impressions persisted to some degree until the late 1940s and early 1950s, when they were clearly shown to be in error. In the interim, however, these misconceptions were responsible for setting back acceptance of the concept that DNA of chromosomes carried genetic information.

Although some forms of cellular DNA exist as **single-stranded structures,** the most widespread DNA structure is the **double helix.** The double helix can be visualized as resulting from interwinding of two right-handed helical polynucleotide strands around a common axis. The two strands achieve contact through hydrogen bonds, which are formed at the hydrophilic edges of their bases. These bonds extend between purine residues in one strand and pyrimidine residues in the other, so that the two types of resulting pairs are always adenine-thymine and guanine-cytosine. A direct consequence of these hydrogen-bonding specificities is that double-stranded DNA contains equal amounts of purines and pyrimidines. Examination of space-filling models clearly indicates structural compatibility of these bases in forming linear hydrogen bonds.

This relationship between bases in the double helix is described as **complementarity.** Bases are complementary because every base of one strand is matched by a complementary hydrogen-bonding base on the other strand. For instance, for each adenine projecting toward the common axis of the double helix, a thymine must be projected from the opposite chain so as to fill exactly the space between strands by hydrogen bonding with adenine. Neither cytosine nor guanine fits precisely in the available space across from adenine in a manner that allows formation of hydrogen bonds across strands. These **hydrogen-bonding specificities** ensure that the entire base sequence of one strand is complementary to that of the other strand.

The double helix exists in various geometries designated as **DNA A, B, and C.** Formation of these different **conformations** depends on the base composition of DNA and on physical conditions. These forms share certain common characteristics. Specifically, the phosphate backbones are always located on the outside of the helix. Also, because diesters of phosphoric acid are fully ionized at neutral pH, the exterior of the helix is negatively charged. Bases are well packed in the interior of the helix, where their faces are protected from contact with water. In this environment the strength of hydrogen bonds that connect bases can be maximized. Interwinding of two strands produces a structure having two helical grooves that separate the winding phosphate backbone ridge.

However, the precise **geometry of the double helxi** varies among the different forms.

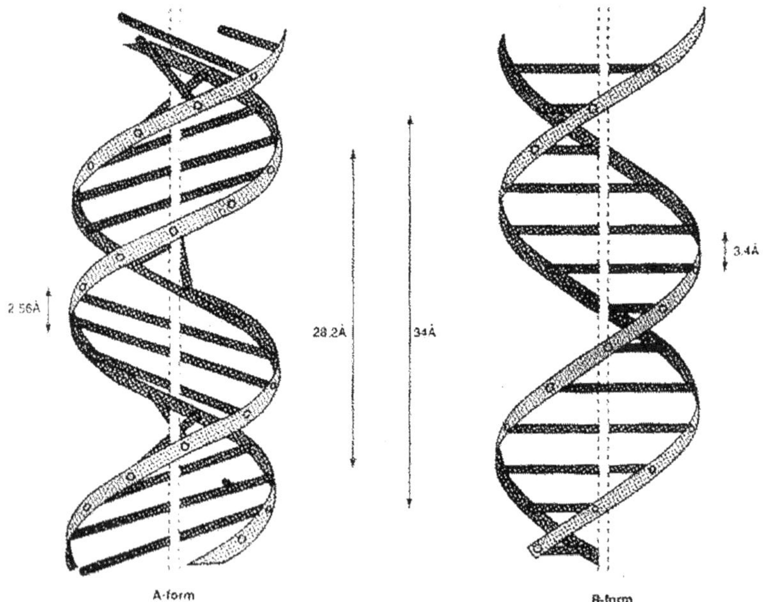

Various geometries of DNA double helix.

The original X-ray data obtained with highly oriented DNA fibers suggested occurrence of a form, later designated as B, which appears to be that commonly found in solution and *in vivo*. A characteristic of this form is that one of its grooves is wider (**major groove**) than the other **(minor groove).** Disparity in width between these two grooves results from the characteristic geometry of base pairs (bp). Glycosidic bonds between sugars and bases of each base pair are not arranged directly opposite to one another. Instead the edge of the helix, that is more than 180° from glycosidic bond to glycosidic bond, is the edge that forms part of the major groove. Clearly, the opposite edge corresponds to the minor groove. The nucleotide sequence of a polynucleotides can be discerned without dissociating the double helix by looking inside these grooves. As each of the four bases has its own orientation with respect to the rest of the helix, each base always shows the same atoms through the grooves. C-6, N-7, and C-8 of the purine rings and C-4, C-5, and C-6 of the pyrimidine rings line up in the major groove. The minor groove is paved with C-2 and N-3 of the purine and C-2 of the pyrimidine rings. Forms A and C differ from B in the pitch of the base pairs relative to the helix axis, as well as in other geometric parameters of the double helix, including conformation of sugar residues, which is one of the more flexible components of the DNA molecule. Alternative forms of the double helix are the result of **conformational variations** of the sugar-phosphate groups that form the backbone of constituent polynucleotides. The conformation of the furanose ring of sugar residues exists in nonplanar (puckered) forms. This ring may be visualized in the form of an envelope with four carbon atoms at the corner of the envelope. Oxygen is positioned at the top of the envelope flap and therefore may bend out of the envelope body.

The main body of the envelope may also be twisted. Twisting the C-2′ and the C-3′ atoms relative to the other atoms produces two distinct forms. C-2′ twists up from the plane and results in the C-2′ **endo** form. As a rule, atoms that are positioned on the same side of the plane as C-5′ have by definition the **endo conformation.** The C-2′ **endo** and C-3′ **endo** are the most common conformers fo[illegible]eic acids, while free nucleotides in solution are characterized by a rapid equilibrium between these conformers. Another variation in nucleic acid conformations arises from rotations about the C-1′–*N*-glycosidic bond that is responsible for variants known as **syn** and **anti** forms. **Anti** conformations are the predominant forms in nucleic acids, while in free nucleotides in solution **syn-anti** equilibrium depends on the nature of the base. Generally, purine nucleotides are characterized by a rapid **syn-anti** equilibrium while pyrimidines usually adopt **anti** conformations.

Finally, conformational variations in DNA may result from relative orientations of the planes of the bases between strands. Differences in orientation between planes of H-bonded bases may produce double helix variants with different base **tilt, roll, twist,** or **propeller twist.** For example, DNA forms A and B differ drastically in base **tilt** and deviations of **tilt** and **roll** angles, occurring in phage tracts of adenine residues, are responsible for extensive bending of the double helix axis over certain functionally important regions of DNA. Under conditions of low salt concentration and humidity, the thin B-DNA double helix shifts to a conformation characterized by a thicker helix. In this conformation nucleotides move off center toward the major edge of each base pair, generating A-DNA, which has a narrower and deeper major groove and a wider and shallower minor groove than B-DNA. The parameters for these different DNA conformations, have been determined of DNA by X-ray diffraction methods. While the numbers provide very accurate information about molecular geometry and dimensions of crystalline samples, they give only average dimensions for monomeric units present in a noncrystalline macromolecule. Therefore these parameters are listed as such and the listing does not imply that the same geometry characterizes each and every individual base pair in DNA. Rather, depending on base sequence, considerable local variation in conformation of individual nucleotides may occur.

An important structural characteristic of double-stranded DNA is that its strands are **antiparallel.** Polynucleotides are asymmetric structures with an intrinsic sense of **polarity** built into them. The two strands are aligend in opposite directions; if two adjacent bases in the same strand, for example, thymine and cytosine, are connected in the 5′→3′ direction, their complementary bases adenine and guanine will be linked in the 3′→5′ direction (directions are defined by linking the 3′ and 5′ positions within the same nucleotide). This antiparallel alignment produces a stable association between strands to the exclusion of the alternate parallel arrangement.

The double-stranded structure of DNA was proposed in 1953, partly based on previously available X-ray diffraction studies suggesting that the structures of DNA from various sources exhibited remarkable similarities. These studies also suggested that DNA had a helical structure containing two or more polynucleotides. Evidence of central importance to the proposal was the clarification of the quantitative base composition of DNA, indicating the molar equivalence between purines and pyrimidines essential for the complementarity between the two strands.

9.3 Denaturation and Renaturation

The double helix is disrupted during almost every important biological transformation in which DNA participates, including DNA replication, transcription, repair, and recombination. Therefore the forces that hold the two strands together are adequate for providing stability and yet weak enough to allow facile strand separation. In fact, the double helix is stabilized relative to the single strands by about 1 kcal per base pair. Therefore a relatively minor perturbation can produce disruption in double strandedness, provided that only a short section of the DNA is involved. As soon as the relatively few base pairs have separated, they close up again and release free energy, and then the adjacent base pairs unwind. In this manner minor disruptions of double strandedness can be propagated along the length of the double helix. Thus, at any particular moment, the large majority of the bases of the double helix remain hydrogen bonded, but all bases can pass through the single-stranded state, a few at a time. This **dynamic state** of the double helix is characterized by the movement of an "open-stranded" portion up and down the length of the helix. The "dynamic" nature of this structure is an essential prerequisite for the biological functions of DNA as it undergoes repair or recombination.

Separation of DNA strands can be studied by increasing the temperature in solution. At relatively low temperatures a few base pairs will be disrutped, creating one or more "open-stranded bubbles". These "bubbles" form initially in sections that contain relatively higher proportions of adenine and thymine pairs because of the lower stacking energies of dimers of such pairs. As the temperature is raised, the size of the "bubbles" increases and eventually the thermal motion of the polynucleotides overcomes the forces that stabilize the double helix. At even higher temperatures the strands separate physically and acquire a randon-coil conformation. The process is most appropriately described as a **helix-to-coil transition,** but it is commonly called **denaturation.** This is accompanied by a number of physical changes, including a buoyant density increase, reduction in viscosity, change in ability to rotate polarized light, and changes in absorbancy.

Changes in absorbance are frequently used to follow the process of denaturation experimentally. DNA absorbs in the UV region due to the heterocyclic aromatic nature of its purine and pyrimidine constituents. Although each base has a unique absorption spectrum, all bases exhibit maxima at or near 260 nm. This property is responsible for the absorption of DNA at 260 nm.

This absorbancy can be as much as 40% lower than that expected from adding up the absorbancy of each of the base components of DNA. This property of DNA, referred to as **hypochromic effect,** results from the stacking of the bases along the DNA helices. In this arrangement, interactions between the π electrons of neighbouring bases produce a decrease in absorbancy.

Now let us try to know about renaturation. Complementary DNA strands separated by denaturation can reform a double helix if appropriately treated. This is called renaturation or reannealing. If denaturation is not complete and only a few bases remain hydrogen bonded between the two strands, the helix-to-coil transition is rapidly reversible.

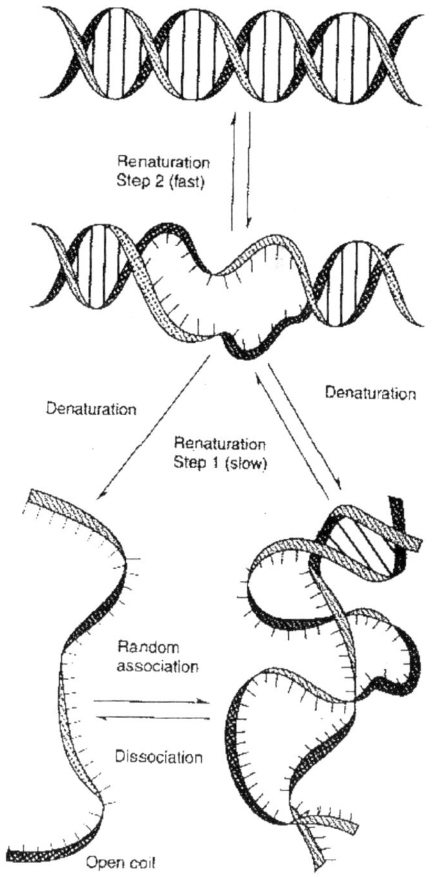

Denaturation of DNA.

Annealing is possible even after the complementary strands have been completely separated. Under these conditions the renaturation process depends on the meeting of complementary DNA strands in an exact manner that can lead to the reformation of the original structure, and it is a slow, concentration-dependent process. As a rule, maintaining DNA at 10-15°C below its T_m, under conditions of moderate ionic strength (0.15 M), provides the maximum opportunity for renaturation. At lower salt concentrations, the charged phosphate groups repel one another and prevent the strands from

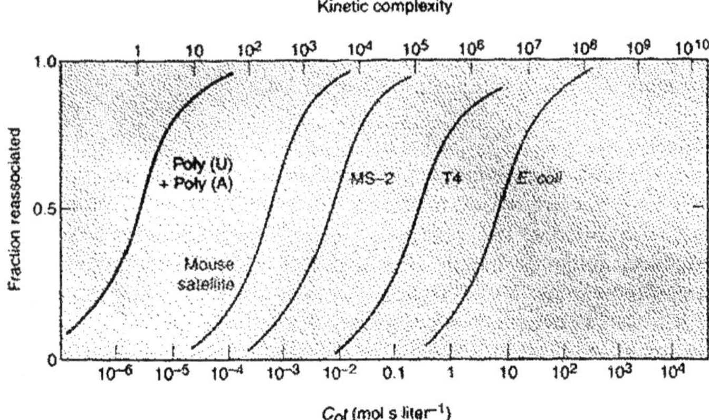

Reassociation kinetics for DNA isolated from various sources.

associating. As renaturation begins, some of the hydrogen bonds formed are extended between short tracts of polynucleotides that might have been distant in the original native structure. Renaturation is facilitated by the presence of short sequences, consisting of four to six base pairs, reiterated many times within every DNA strand. A large number of much longer nucleotide sequences are repeated many times within the eukaryotic genome. Such sequences provide sites for initial base pairing that produces a partially hydrogen-bonded double helix. These randomly base-paired structures are short-lived because the bases that surround the short complementary segments cannot pair and lead to the formation of a stable fully hydrogen-bonded structure. However, once the correct bases begin to pair by chance, the double helix over the entire DNA molecule is rapidly reformed. Renaturation is a two-step process. The first step determines the rate of association, involves the chance meeting of two complementary sequences on different strands, and is therefore a second-order reaction. The rate of renaturation is thus proportional to the product of the concentrations of the two homologous dissociated strands and is expressed as $dt/dc = -kc^2$, where k is the rate constant for the association. Integration of this equation gives $C/C_0 = 1/(1+kC_0t)$, where C is the concentration of single-stranded DNA expressed as moles of nucleotide per liter at time t, and C_0 is the concentration of DNA at time zero. A plot of C/C_0 (which is proportional to DNA that is single stranded or of the DNA fraction that is reassociated) versus C_0t can be constructed, and a $C_0t_{0.5}$ **(Cot-a-half)** value, which corresponds to $C/C_0 = 0.5$ can be determined. The $C_0t_{0.5}$ value is proportional to the complexity of the genome.

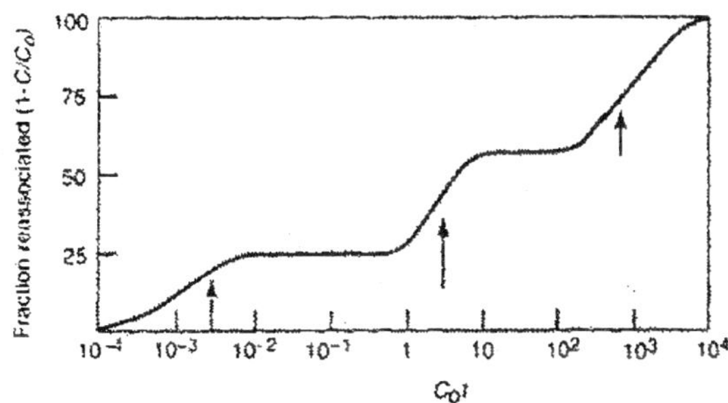

	Fast Component	Intermediate Component	Slow Component
Percent of genome	25	30	45
$C_0t_{1/2}$	0.0013	1.9	630
Complexity, bp	340	6.0×10^5	3.0×10^8
Repetition frequency	500,000	350	1

Reassociation kinetics of eukaryotic DNA.

Complexity is equal to the molecular mass of the genome provided that the genome consists of unique nucleotide sequences. For example, both the complexity and molecular mass of a hypothetical genome consisting of three unique nucleotide sequences that may be represented as N_1, N_2, and N_3 is equal to the sum $N_1 + N_2 + N_3$. However, in eukaryotic genomes, which contain both unique as well as reiterated sequences, the complexity of the genome is significantly lower than the molecular mass. If, for instance, a eukaryotic genome contains 10^5 copies of sequence N_3, 10^3 copies of sequence N_2, and 1 copy of sequence N_1, the complexity will still be $N_1 + N_2 + N_5$ but the molecular mass will be equal to 10^5N_3

+ 10^3N_2 + N_1. Thus complexity may be defined as the minimum length of DNA that contains a single complete copy of all the single and **reiterated sequences** that are represented within the genome.

The C_0t curves of eukaryotic genomes with reiterated DNA segments show several kinetic components, each representing those parts that have similar reiteration frequencies. Highly reiterated sequences will reassociate the fastest; unique sequences are the slowest. Thus C_0t curves provide information on **genome complexity,** on the number of **repetitive classes,** and on the proportion of the total genome represented by those classes.

9.4 Determination of DNA Size

The length of DNA varies from a few thousand base pairs for DNA of the small viruses, to millions for chromosomal DNA of bacteria, and to billions for the chromosomal DNA of animals. DNA sizes can be expressed as number of base pairs, molecular mass, the length of the strands, and even the actual mass of DNA. The units used in these expressions, however, can easily be interconverted, since a DNA of mol wt 1×10^6 contains approximately 1500 bp and is 0.5 nm long. DNA mass can be converted to molecular mass by division with the average molecular mass of a DNA nucleotide pair.

The amount of DNA per cell increases as the complexity of the cellular function increases. Although mammalian cells contain some of the highest amounts of DNA per cell, some amphibian, fish, and plant cells may contain even higher amounts. In fact, lung fish cells contain more than 40 times the amount of DNA in human cells, but such extraordinary amounts of DNA reflect a **reiteration of nucleotide sequences** within the DNA macromolecule and do not represent an actual increase in the size of DNA in terms of unique sequences, that is, **DNA complexity.** The size of the DNA of higher cells is very large indeed. The DNA contained within a single human cell is packaged in the form of 46 chromatin fibers or chromosomes. In its most condensed state, that is, during metaphase, the largest of these chromosomes is about 10 μm. If the DNA packaged within this chromosome were stretched out in the conventional B-DNA form, it would be over 8 cm long, that is, 8000 times longer than it is when packed within the chromosome. This suggests that the polynucleotides are exquisitely packed in order to fit within the minute dimensions of the cell nucleus.

Because of their extraordinary length, relative to the total mass, DNA molecules are extremely sensitive to shearing forces that develop during ordinary laboratory manipulations. Even careful pipetting may shear a DNA molecule. During the process of isolation it is difficult to prevent with absolute confidence the disruption of some phosphodiester bonds by contaminating endonucleases (nicking). For these reasons the precise size of DNA of higher species could not be determined until special handling techniques were developed, both for the isolation of DNA and the measurement of its molecular mass.

DNA Cell Content of Some Species

Types of Cell	*Organism*	*DNA per cell (pg)*[a]
Phage	T4	2.4×10^{-4}
Bacterium	*E. coli*	4.4×10^{-3}
Fungus	*N. crassa*	1.7×10^{-2}
Avian erythrocyte	Chicken	2.5
Mammalian leukocyte	Human	3.4

Classical methods for determining size in proteins proved to be unsuitable for measuring the molecular mass of even relatively small DNAs. Custom-tailored methods were devised. **Equilibrium centrifugation** in a density gradient (usually a concentrated cesium chloride solution), **electron microscopy**, and **electrophoresis** in agarose gels are among the principal methods providing reliable information about the molecular masses of DNAs. Electron microscopy provides a measure of the length of DNA strands. Molecular masses can be calculated from known values of the mass per unit length. The DNA can be visualized under the electron microscope if it is first coated with protein and a metal film. Determination of molecular masses by electrophoresis depends on the molecular sieving effect of porous agarose gels. Over a limited range of molecular masses the mobility of DNA is directly proportional to the logarithm of the molecule's weight.

To determine the molecular mass of DNA by equilibrium centrifugation a small portion of a DNA solution to be analyzed is layered on top of a gradient in a centrifuge tube. Upon centrifugation, the molecules of DNA sediment to equilibrium through the gradient. Under these conditions a homogeneous high molecular mass DNA will form a Gaussian band centered at a position in the gradient that corresponds to the density of the DNA. Molecules with different densities are resolved into a series of bands that sediment independently of one another. A relationship can be demonstrated between the width of the bands at equilibrium and the molecular masses.

Labeling of the terminals of DNA has been used successfully for determining molecular masses. DNA is treated with the enzyme alkaline phosphate, which converts the 5′-phosphate nucleotide terminals of double-stranded DNA to the corresponding OH groups. These terminals are then esterified, using [γ-^{32}P]ATP with the enzyme polynucleotide kinase. The free 5′ terminus of each polynucleotide chain becomes labeled. The labeled DNA is then analyzed by **zonal centrifugation** and detected from both its absorbancy at 260 nm and ^{32}P content. The molecular mass is

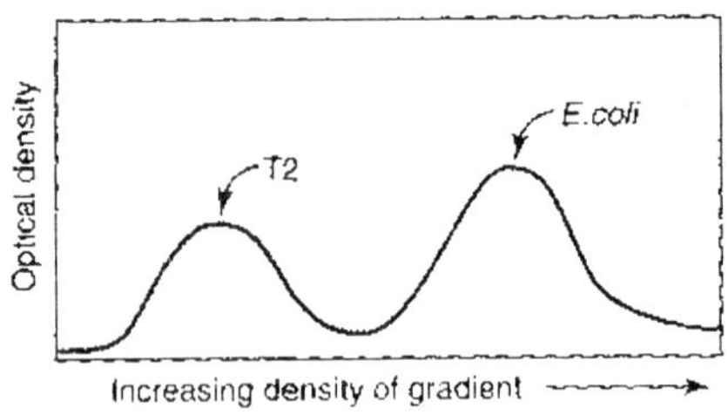

Equilibrium gradient centrifugation of DNA.

calculated from the ratio of the amount of ^{32}P to the absorbancy, both measured at the coinciding peaks of the bands.

Double-stranded circular DNA, with few exceptions, has an intriguing **topology.** The circular structure contains twists, referred to as **supercoils,** which are visualized by electron microscopy. In principle, linear DNA could be converted to a circular molecule. Circular DNA may be formed by bringing together, and joining by a phosphodiester bond, the free terminals of linear DNA. If no other manipulations are introduced, the resulting circular DNA will be **relaxed;** that is, it will have a thermodynamically favoured structure of the linear double helix (B-DNA), which accommodates one complete turn of the helix for approximately 10 base pairs. However, if before sealing the circle, one DNA terminus is held steady while the other terminus is rotated in a direction that unwinds the double helix, the resulting structure will be strained. This strained structure, which is characterized by a deficit of turns, is known as **negative superhelical DNA.** Negatively supercoiled DNA is underwound in that it has fewer helical turns than what the molecule would accommodate as a linear or as a relaxed structure. The underwinding results in participation of more base pairs per helical turn, which produces a decrease in the angle of twist between adjacent base pairs. Therefore underwinding generates **torsional tension.** Torsional strain increases the standard free energy of DNA by about 10 kcal mol^{-1} per each supercoil that is introduced into the structure. The strain produced by this deficit of turns is accommodated by the disruption of hydrogen bonds and the opening of the double helix over a small region of the macromolecular structure. The resulting structure may be viewed as consisting of a small-stranded loop along with regions of regularly spaced double helical turns with a geometry similar to relaxed B-DNA. If, however, hydrogen bonds are not disrupted, the circular DNA will twist in a direction opposite to the one in which it was rotated initially in order to relieve the strain induced by the unwinding. Thus the rotational strain that was introduced before the circularization of DNA can be accommodated either by the disruption of H bonding or by the formation of tertiary structures with visible supercoils. These two representations of the negative superhelix should be viewed as two manifestations of the same underlying phenomenon. In general, a dynamically imposed compromise, determined by the environment and the status of circular DNA, is reached between hydrogen-bond disruption and supertwisting. In practice, this means that supercoiled DNA may consist of twisted structures with enhanced tendency to generate regions with disrupted hydrogen bonding (bubbles).

9.5 Superhelical DNA

Various forms of supercoiled DNAs can be described using *L, T,* and *W* numbers. It should be recalled that the turns of the typical double helix are right handed. Therefore, if a hypothetical linear DNA duplex that is 10 turns long (L = 10 and T = 10) is unwound by, say, one turn, the resulting strucutre will have the following characteristics: L = 9 and T = 9. A potentially equivalent structure can be formed if instead ends of the same hypothetical DNA are secured so that they cannot rotate and the molecule is looped in a counterclockwise manner. Since in this case untwisting is not permitted to occur, the number of helical turns remains

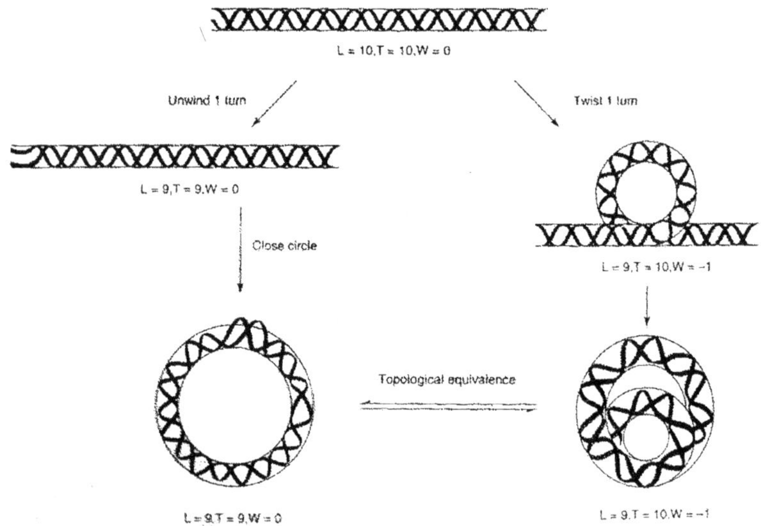

Various types of DNA superhelices.

unchanged; that is, $T = 10$. However, as a result of "looping" operations, linking number is now reduced by 1; that is, $L = 9$. The structure resulting from this deliberate introduction of a loop is visibly superhelical. Furthermore, application of the equation that relates values of *L, T,* and *W* indicates that *W* must be equal to –1; that is, the structure is a negative superhelix with one supertwist.

The two structures described abvoe —$L = 9$, $T = 9$, $W = 0$ and $L = 9$, $T = 10$, and $W = -1$—obviously have the same linking number and are therefore interconvertible without the disruption of any phosphodiester bonds. The potential equivalence of these two types of structure becomes more apparent when ends of polynucleotides in each structure are joined into a circle without strands being allowed to rotate. Circularization produces an **interwound** circular structure (a number 8-shaped structure referred to as a **plectonemic** coil) or a doughnut-shaped superhelical arrangement referred to as a **toroidal turn,** both of which are freely interconvertible. An interwound turn, can be produced by unfolding a toroidal turn along an axis that is distinct from the supercoil axis.

In summary, if the termini of a linear DNA molecule are covalently attached, a "relaxed" covalent circle results. However, if one end of the double helix is maintained in a fixed and stationary position while the other end is rotated in either direction prior to closing the circle, the resulting structure will twist in the opposite direction so as to generate a supertwisted helical structure. For each additional complete turn of the helix, DNA will acquire one more superhelical twist in the opposite direction of rotation in order to relieve intensifying strain. As a result, topologically equivalent structures, will be created. A real superhelical DNA exists as an equilibrium among these forms and many other intermediate arrangements in space that have the same linking number but different numbers of helical turns and supertwists. Although linking number is a constant and an integer, the number of twists can change in positive and negative increments, which are compensated by negative and positive changes in the writhing number. DNA supercoils are distributed in part as mixtures of interwound (plectonemic) and toroidal coils and as decreases in twist angle of the double helix. The interwound form is by far the more predominant structure for supercoiled DNA. In solution about 70% of the deficiency in linking numbers may be distributed as writhe change and 30% as changes in twist.

Although the closed circular form of DNA is an ideal candidate for acquiring a superhelical structure, any segment of double-stranded DNA that is in some way immobilized at both of its terminals qualifies for superhelicity. This property therefore is not the exclusive province of circular DNA. Rather, any appropriately anchored linear DNA molecule can acquire a superhelical conformation. The DNA of animal cells, for instance, normally associated with nuclear proteins, falls into this category. Animal DNA can acquire a superhelical form because its association with nuclear proteins creates numerous closed topological domains. A topological domain is defined as a DNA segment contained in a manner that restrains rotation of the double helix. In addition, circular DNAs of most bacterial phages, animal viruses, bacterial plasmids, and cell organelles, such as mitochondria and chloroplasts, contain superhelical DNAs.

9.6 Slipped, Mispaired DNA

DNA regions with direct repeat symmetry can form structures known as **slipped, mispaired DNA** (SMP-DNA). Their formation invovles the unwinding of the double helix and realignment and subsequent pairing of one copy of the direct repeat with an adjacent copy on the other strand. This realignment generates a single-standed loop. Two isomeric structures of a SMP-DNA are possible. One generates a loop consisting of the 5′ direct repeat in both strands and the other produces loops of the 3′ direct repeat. Although SMP-DNA has not yet been identified, genetic evidence suggests that this type of DNA is undoubtedly involved in spontaneous frameshift mutagenesis that is manifested as base addition or deletion occurring within runs of single bases. First, a homopolymeric sequence in one strand (template strand) unpairs from a newly synthesized complementary strand and reforms hydrogen bonds with a different set of bases, resulting in the formation of an extrahelical base on either the template strand or progeny strand. Continued replication produces a deletion when the progeny strand slips forward or a duplication when the strand slips backward. Deletions and duplications of DNA segments, longer than a single base, occur during DNA replication between direct repeats, which can form slipped-looped structures. Duplication of certain simple triplet repeats that are implicated as the basis of several human genetic diseases may also occur by this mechanism.

DNA in eukaryotic cells is associated with various types of protein to form **chromatin.** In resting (nondividing) cells, chromatin is amorphous and dispersed within the nucleus. Just prior to cell division (mitosis), chromatin becomes organized into compact structures (fibers) called **chromosomes.**

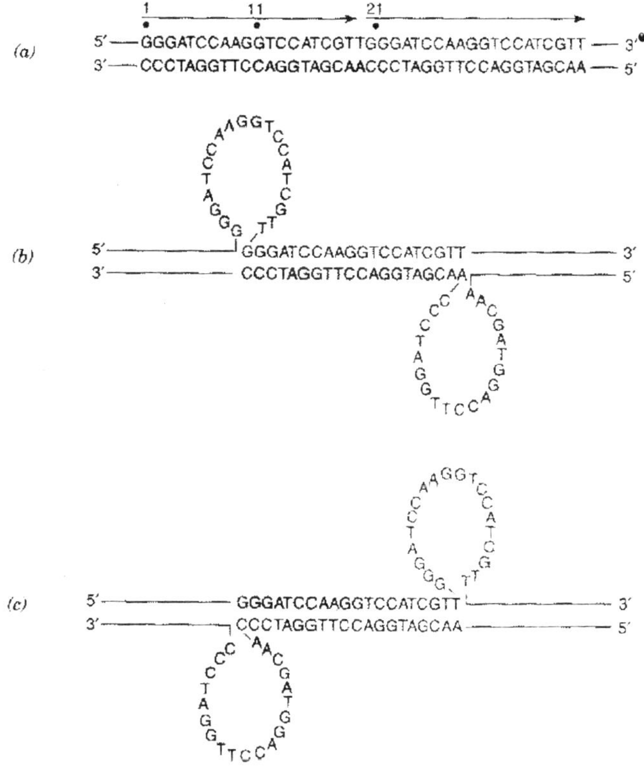

Slipped, mispaired DNA.

The division of genetic information into numerous independent domains, that is, chromosomes, may be necessitated by the enormous length of the genome of most eukaryotes. Each chromosome is characterized by the presence of a **centromere,** which functions as a site for attachment to proteins that link the chromosome to the mitotic spindle. Sister chromatids are connected at the centromere. **Telomeres** define the termini of linear chromosomes. A third element that characterizes chromosomes is the presence of a sequence required for the initiation of DNA replication **(origin of replication).** The number of chromosomes observed is species specific with human cells containing 46 chromosomes (chromatids) organized into 23 pairs. The average DNA length of each one of these chromosomes is 1.3×10^8 nucleotide pairs or approximately 5 cm. It is believed that each human chromosome consists of a single intact DNA molecule varying in size from 263×10^6 base pairs for chromosome 1 to less than 50×10^6 bp for chromosome 23. If the DNA of all 46 chromosomes were lined up in the B-DNA conformation, it would be more than 2 m long.

The chromosomal organization that makes it possible for DNA to fit within a cell nucleus with a diameter of approximately 5 μm requires a "condensation ratio" of more than five orders of magnitude. During metaphase the DNA molecule is very tightly wound. For example, human chromosome 16 is 2.5 μm long, whereas the DNA molecule is 3.7 cm in each of the two chromatids, giving a condensation ratio of 1.5×10^4:1. The parceling of DNA in 46 chromosomes provides for a further increase in the condensation ratio to 10^5:1. This remarkable degree of condensation of cellular DNA. The early stages of DNA packing that lead to formation of 30-nm fibers have been extensivley studied. The latter stages, in which looped domains of the 30-nm fiber are organized into scaffolds and chromatid coils, are based on indirect evidence and are more speculative. At each stage of packing, shown in this model, DNA is condensed severalfold. The cumulative effect of the successive folding stages provides the large condensation ratio necessary for the **packing of DNA** within the nucleus. The first stage of organization is the formation of a **"beads-on-a-string"** structure consisting of DNA associated with a class of highly basic proteins known as **histones.** These bind tightly to DNA, forming very stable complexes. The "beads-on-a-string" arrangement is seen in chromatin treated under conditions of low ionic strength and examined under the electron microscope.

9.7 Nucleosomes

Histones interacting with DNA form the periodic "beads-on-a-string" structure, called a **polynucleosome,** in which an elementary unit, a nucleosome, is regularly repeated. Each **nucleosome** is a disk-shaped structure about 11 nm in diameter and 16 nm in height that consists of a DNA segment and a histone cluster composed of two molecules each of H2A, H2B, H3, and H4 histones. The clusters are organized as tetramers consisting of $(H3)_2$-$(H4)_2$ with an H2A-H2B dimer stacked on each face in the disk. The DNA is wrapped around the octamer as a negative toroidal superhelix at a pitch of about 30 Å with the central $(H3)_2$-$(H4)_2$ core interacting with the central 70-80 bp of the DNA wrap. Histones are in contact with the minor groove of DNA and leave the major groove available for interaction with the proteins that regulate gene expression and other DNA functions. Two distinct structures of nucleosomes can be distinguished: the **nucleosome core** and the **chromatosome.** The chromatosome constitutes the most elementary structural unit of nucleoproteins. These two structures are obtained by the digestion of polynucleosomes with nucleases (DNases) that, depending on conditions, can remove most or all DNA that is not tightly bound with histones. Nucleosomes obtained by nuclease digestion can be crystalized and studied by X-ray diffraction.

Periodicity of distribution of nucleosomes along the poly-nucleosome structure has been determined by controlled digestion with a nuclease that preferentially attacks linker DNA. The digestion pattern suggests the presence of nucleoprotein segments, which on the average contain about 200 bp of DNA or multiples of 200 that result from incomplete digestion. The relationship between size of segments and expected number of nucleosomes associated with them has been confirmed by electron microscopy.

With the exception of a small amount of eukaryotic DNA, which is located in mitochondria and chloroplasts and which occurs in the form of small superhelices generally free of protein, all eukaryotic DNA is associated with histones.

Although nucleosomes are periodically positioned along the polynucleosome, their distribution is not random with respect to the base sequence of DNA. DNA does not bend uniformly but rather bends gently and then more sharply around the histone octamers. This suggests that DNA binding is sequence dependent and that **nucleosome positioning** may be influenced by the nucleotide sequence of DNA. In fact, nucleosomes tend to associate preferentially with certain DNA regions. DNA tracts that resist binding, such as long A tracts or G-C repeats, are not usually associated with nucleosomes. In contrast, certain bend DNA regions, for instance, periodically phased A tracts, assoicate strongly with histones. The majority of nucleosome core particles can relocate over a cluster of positions along the DNA separated by about 10 bp. The resulting mobility of these coil particles probably allows DNA polymerases and other enzymes to gain access to specific DNA sequences. The organization of DNA into nucleosomes appears to have fundamental consequences for transcription and DNA repair.

How polynucleosomes are organized into higher structures is not fully understood. Models as to the higher levels of packing of 30-nm fibers are based on indirect evidence obtained from studies of two specialized types of chromosomes—the **lampbrush chromosomes** of vertebrate oocytes and **the polytene chromosomes** of fruitfly giant secretory cells. These chromosomes are exceptional in that they maintain precisely defined higher-order structures in interphase, that is, when cells are in a resting (nondividing) state. The structural features of interphase lampbrush chromosomes have led, by extrapolation, to the proposal that chromosomes in general are organized as a series of looped, condensed domains of 30-nm fibers of variable size for different organisms.

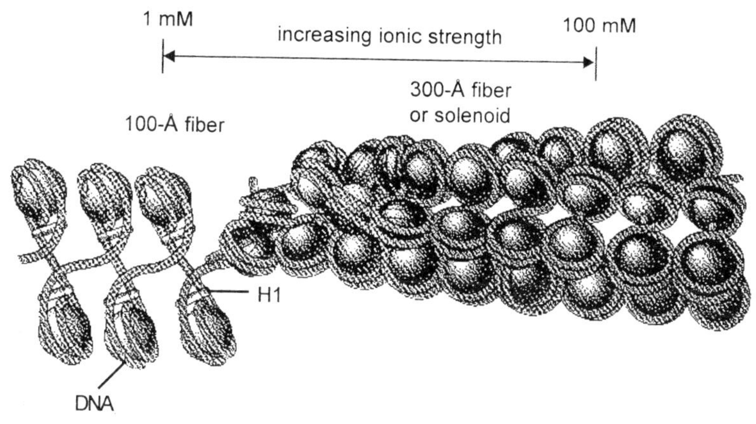

Nucleofilament structure.

It is estimated that these loops may contain anywhere from 5000 to 120,000 bp with an average of about 20,000. Thus the haploid human genome of 3×10^9 bp would correspond to about 60,000 loops, which is close to the estimated number of genes of 70,000 to 100,000. It appears likely that each loop contains one or a few linked genes. The domains are bound to a nuclear scaffold consisting of H1 histone and several nonhistone proteins, including two major **scaffold proteins Sc1** (a topoisomerase II) and **Sc2.** The loops are fixed at their bases and therefore they can accumulate supercoils. Specific AT-rich regions of DNA known as **SARs (scaffold attachment regions)** are preferentially associated with the scaffold. SARs also contain topoisomerase II binding sites. The presence of type II topoisomerase at the base of closed topological domains, which define the scaffold loops, suggests that supercoilding and supercoiling changes within these domains are biologically important functions. Formation of looped domains may account for as much as an additional 200-fold condensation in the length of DNA and an overall packing ratio of more than four orders of magnitude. Each loop can be coiled and then supercoiled into 0.4 μm of a 30-nm fiber. Since the thickness of a sister chromatid is about 1 μm in diameter, packing of the 20-nm fiber into a chromatid would require just one more order of folding.

9.8 Endonucleases

Restriction Endonucleases cleave DNA chains at a specific sequence, making possible the sectioning of large DNA molecules into small segments. These highly specific bacterial enzymes act by making two cuts one in each strand of double-stranded DNA of an invading phage generating 3′-OH and 5′-p termini.

This fragmentation exposes phage DNA to eventual degradation by bacterial exonucleases. The terminology for these endonucleases originates from the bacterial sources from which they are isolated. The first three letters of the name is an abbreviation of the species from which the enzyme is isolated. The next letter (or letters) designates the strain of the source and the Roman numeral simply refers to the order in which the enzyme was discovered from the strain. Many hundreds of restriction endonucleases have been isolated in pure form and the list of new restriction enzymes is growing daily. With few exceptions, these enzymes have been found to recognize sequences four to six nucleotides long. These sequences are completely **symmetrical inverted repeats,** known as **palindromes**. The order of the bases is the same when the two strands of the palindrome are read in opposite directions. For example, in the case of the restriction enzyme EcoR1, isolated from *E. coli,* the order of the bases is GAATTC when read from the 5′ terminus of either of the strands.

Restriction endonucleases are classified into three categories. Types I and III make cuts in the vicinity of the recognition site in a unpredictable manner. Type II specifically cleaves DNA within the recognition sequence.

These enzymes recognize specific sequences that occur along large DNAs with relatively low frequencies and fragment DNA very selectively. For example, a typical bacterial DNA, which may contain about 3×10^6 bp, will be cleaved into a few hundred fragments. A small virus or plasmid may have few or no cutting sites at all for a particular restriction endonuclease. The practical significance of this selectivity of restriction enzymes is that a particular enzyme generates a unique family of fragments for any given DNA molecule. This unique fragmentation pattern is called a **restriction digest.**

EcoRI	5'.... GAATTC .... 3' 3'.... CTTAAG .... 5'	→	5'.... G3' 3'.... CTTAA5'	+ 5'AATTC .... 3' 3'G .... 5'
PstI	5'.... CTGCAG .... 3' 3'.... GACGTC .... 5'	→	5'.... CTGCA3' 3'.... G5'	+ 5'G .... 3' 3'ACGTC .... 5'
HaeIII	5'.... GGCC .... 3' 3'.... CCGG .... 5'	→	5'.... GG3' 3'.... CC5'	+ 5'CC ....3' 3'GG ....5'

Types of products generated by type II restriction endonucleases.

The availability of restriction enzymes for sectioning large DNA sequences and the development of new gel electrophoresis techniques for separating DNA segments have made the determination of sequences a simple matter.

Early attempts to determine DNA sequences were limited to small DNA fragments that could easily be separated from the remaining DNA. Sequences that bind selectively with various functional proteins, for example, RNA polymerase and the repressor proteins, were among the first to be determined. The binding protein protects the DNA section over which it is bound from the action of a nuclease and the protected DNA is recovered after digestion and removal of the protein. These studies indicated that many functional proteins and enzymes interact with DNA over regions of palindromic sequence.

Palindromes in DNA also serve as recognition sites for **methylases** that modify the host DNA by introducing methyl groups into two bases of the palindrome. Once methylated, these palindromes cannot be recognized by the corresponding restriction enzymes, and the DNA of the host is protected from cleavage.

Contemporary sequencing methods have made possible determination of the complete nucleotide sequences of the DNA of viruses and small bacteria and the partial sequence of many eukaryotic genomes. An ambitious current goal of DNA sequencing is the determination of the sequence of the entire human genome, which consists of almost 3×10^9 bp, and that of several other mammalian organisms.

9.9 Mutations

Mutations are stable changes in DNA structure. One of the fundamental requirements for a structure that serves as a permanent depository of genetic information is high stability. Such stability is essential, at least in those parts that code for the genetic information. The structure of the DNA bases, however, is not totally exempt from gradual change. Normally, changes occur infrequently and they affect very few bases. Chemical and irradiation-induced reactions modify the structure of some bases, disrupt phosphodiester bonds, and sever strands. Extensive chemical changes of the bases occur spontaneously. Errors also occur during replication and DNA recombination, leading to incorporation of one or more erroneous bases. In almost every instance, however, a few cycles of DNA replication are required before a modification in the structure of a base can lead to irreversible damage. In effect, DNA polymerases must use the polynucleotide initially damaged as a template for the synthesis of a complementary strand for the initial change to become permanent. Use of the damaged strand as template extends the damage from a change of a single base to a change of a complete base pair and subsequent replication perpetuates the change. Other sources of permanent modifications of DNA include changes resulting from insertion to deletion from a DNA of short or longer nucleotide sequences during the process of DNA recombination. Intercalation of certain planar organic ring structures can also lead to insertion of nucleotides. Finally, deletions may occur as a result of chemical modification of the bases.

Mutations introduced on a DNA strand, such as the replacement of a cytosine by a uracil resulting from deamination of cytosine, extend to both strands when the damaged strand is used as a template during replication. In the first round of replication uracil selects adenine as complementary base. In the second round of replication uracil is replaced by thymine. Similar events occur when the other bases are altered.

Irreversible alteration of a few DNA base pairs can cause drastic changes in the organism. These changes, referred to as **mutations,** may be hidden or visible, that is, **phenotypically silent** or **expressed.** Therefore a mutation is defined as a stable change in the DNA structure of a gene, which may be expressed as a phenotypic change in the organism. Mutations may be classified into two categories: **base substitutions** and **frameshift mutations.** Base substitutions include **transitions,** substitutions of one purine-pyrimidine pair by another, and **transversions,** substitutions of a purine-pyrimidine pair by a pyrimidine-purine pair. Frameshift mutations, which are the most radical, are the result of either the insertion of a new base pair or the deletion of a base pair or a block of base pairs from the DNA base sequence of the gene.

Ultraviolet light, including sunlight, and X-ray irradiation are also effective means of producing mutations. **Radiation energy** absorbed by the DNA induces the formation of minor amounts of the **ionized forms of the bases.** These ionized forms cannot pair with the normal partners of the base, but, instead, they engage in atypical base pairing. The presence of ionized base forms at the moment of DNA replication is therefore expected to increase the frequency of mutation in the newly synthesized DNA strands. UV irradiation of DNA causes formation of dimers between adjacent pyrimidine bases. Activation of the ethylene bond of these bases frequently leads to a **photochemical dimerization** of two adjacent pyrimidines. Thymine residues are particularly susceptible to this reaction, although cytosine dimers and thymine-cytosine combinations are also produced.

High-energy radiation (X-rays or gamma rays) brings about direct modifications in the structure of the bases. Intermediates produced by electron expulsion can be rearranged, leading to the opening of the heterocyclic rings of the bases and the disruption of phosphodiester bonds. In the presence of oxygen additional reactions take place, yielding a variety of oxidation products.

With the appropriate deoxyribonucleoside triphosphates, DNA polymerases function with a high degree of fidelity. Some mutations do occur during DNA replication, but these changes are limited by the high synthetic fidelity of DNA polymerase and the "proofreading" exonuclease properties of this enzyme. The fidelity of DNA replication is further enhanced postreplicatively by an excision repair process known as the **mismatched repair system.** This system recognizes and corrects mismatches in newly replicated DNA by detecting distortions on the outside of the helix that are produced from poor fit between paired noncomplementary bases. Clearly, accurate correction of mismatched bases requires that the mismatched repair system discriminate between preexisting and newly synthesized DNA strands. Such discrimination is feasible because certain adenine residues in DNA, which are part of a recurring GATC sequence, are subject to methylation that occurs post-transcriptionally, but with some delay. Mismatched proofreading is carried out by a multienzyme complex that excises mismatched nucleotides only from newly synthesized strands. The complex identifies these nucleotides by searching for unmethylated adenine residues in the GATC sequences of each strand.

DNA polymerases are unable to distinguish between the normal deoxyribonucleoside triphosphate substrates and other nucleotides with very similar structures, thus leading to their incorporation and a mutation. Classic examples of such analogs are deoxyribonucleotides of 5-bromouracil (5-BrdU) and 2-aminopurine (2-AP) that have been used experimentally for the introduction of mutations. Incorporation of 5-BrdU into DNA introduces, with a high frequency, a transition mutation in which a pu-py pair is transformed to another pu-py. Specifically, 5-BrdU paired with A is changed to a C-G pair, which amounts to a TA $\rightarrow$ GC transition. The unusual pairing properties of 5-BrdU appear to relate to the higher tendency of this base to be transformed to an ionized form, relative to T for which it is a substitute. This occurs presumably because of the higher electronegative nature of the bromine atom in comparison to the corresponding methyl group in thymine.

9.10 The Process of DNA Repair

DNA is the only macromolecule that is repaired rather than degraded. The repair processes are very efficient with fewer than 1 out of 1000 accidental changes resulting in mutations. The rest are corrected through various processes of DNA repair.

Mutation rates can be estimated using two entirely different approaches, that is, from the frequency with which new mutants arise either in populations, such as fruitflies, or in specific proteins in cells growing in tissue culture. These experiments provide estimates of mutation rates of 1 base pair change per 10^9 base pairs for each cell per each generation. On this basis, for an average-sized protein, which contains about 1000 coding base pairs, a mutation may occur once in 10^6 cell generations.

DNA repair is a high-priority process for maintaining cellular function. Germ cells must be protected against high rates of mutation to preserve the species, and somatic mutation must be controlled in order to avoid uncontrolled cell growth and disease. Unchecked accumulation of damage can lead to accumulation of nonfunctional proteins or unregulated growth characteristic of malignant cells.

There are multiple DNA repair pathways and each specializes in a certain type of damage, although some repair pathways have a wider versatility than others. Generally, repair mechanisms are applicable to both prokaryotic and eukaryotic DNA repair.

Repairs may be carried out under rare circumstances as a direct reversal of the damage or, far more commonly, by the replacement of the damaged DNA section. DNA repair depends on the existence of two complementary DNA strands except for **postreplication repair of rare lesions** and **postreplication SOS repair.** Damage or imperfection on one DNA strand can be corrected since the complementary strand provides the necessary information for accurate repairs. Postreplication repair is not a true repair mechanism but rather a stop-gap measure that allows for

DNA replication to occur until damage can be repaired permanently. Postreplication repair cannot use the complementary DNA strand for repairs because this strand is also altered by the replication that precedes the repair. Postreplication repair depends, instead, on another process—DNA recombination. Recombination permits the use of homologous DNA strands, namely, DNA strands with the same or almost the same sequence as the damaged strands, for carrying out the repair of the damaged DNA section. An intriguing feature of DNA repair that has been appreciated recently is its apparent intimate coupling to other central processes in which DNA participates, such as recombination, transcription, and control of the cell cycle. Enzymes involved in DNA repair participate in DNA replication, DNA recombination, and particularly DNA transcription. DNA metabolism integrates important processes that are coordinated through the use of the same molecular tools to achieve different tasks.

Excision repair is catalyzed by different enzymatic systems tailored to specific types of damage. This repair mechanism is universal, occurring in all organisms investigated. The mechanisms are characterized by four sequential steps: incision, excision, resynthesis, and ligation. Incision is the recognition step and is individualized for the specific type of damage present. It is also the rate-controlling step in the process. During excision the damaged DNA section is excised, leaving a gap in the DNA strand. In the resynthesis step the gap is filled by DNA polymerase I. This enzyme functions like DNA polymerase III in that it catalyzes the stepwise addition to nucleotide triphosphates on a 3′-OH generated by the preceding incision step. Polymerase I, however, differs from polymerase III in that it is less processive, tending to dissociate from the DNA after incorporation of 10-12 nucleotides. At this stage the gap is reduced to the size of a single phosphodiester bond. Because of the combined synthetic-nucleolytic action of polymerase I, the nick can move along the strand, undergoing repair until it is finally bridged during the ligation step by the action of DNA ligase. The ligation step appears to be very similar for all types of excision repair.

Base excision repair eliminates modified bases from DNA. The amino groups of cytosine, adenine, and guanine are susceptible to spontaneous elimination, and various chemicals lead to modifications in the structures of purines, including methylation and ring opening. In addition, ring opening may result from exposure to ionizing radiation. Bases that have been deaminated, methylated, or otherwise chemically modified are hydrolytically removed by enzymes referred to as **DNA glycosylases.** This enzyme removes the damaged cytosine, producing a deoxyribose residue with the base missing [**apurinic-apyrimidinic (AP) site**]. AP sites are also generated without the involvement of DNA glycosylases, as in the case of spontaneous hydrolysis of purines (**depurination**) that occurs at very high rates in DNA. AP sites can also result from depyrimidination but the greater stability of the purine-glycoside bond makes this reaction almost insignificant. Once an AP site has been created, the enzyme **AP endonuclease** nicks the phosphodiester backbone at the depurinized site and excises the sugar-phosphate residue. The action of DNA polymerase I and ligase on this structure leads to the restoration of the damaged strand.

A second type of excision repair referred to a **nucleotide excision repair** is activated when DNA is damaged in a way that produces a "bulky" lesion. This occurs when DNA interacts with polycyclic aromatic hydrocarbons, such as benzo[*a*]pyrenes and dialkylbenzathracenes generated by smoking, thymine-psoralene adducts, and guanine-cisplatin adducts formed by chemotherapeutic drugs. UV light-induced dimerization of adjacent pyrimidines also causes bulky lesions. Nucleotide excision repair also corrects other lesions that do not distort the helix, such as the presence of methylated bases. Once the lesion has been located, an endonuclease activity cleaves the modified strand on both sides of the distortion and the entire lesion is removed. Repair is initiated by recognition of the distortion of the DNA by an endonuclease system consisting of the products of three *E. coli* genes *uvrA, uvrB,* and *uvrC.* A tetramer consisting of two **UvrA** and two **UvrB** proteins, which is formed on DNA during a series of preincision steps, "melts" the DNA locally at the expense of ATP and locates the bulky lesion.

The complex is subsequently subjected to incision at both sides of the bulky lesion. First, UvrB makes a 3′ incision and then **UvrC** makes a 5′ incision, leading to the release of an oligonucleotide consisting of 12 or 13 residues that includes the pyrimidine dimer. This nuclease activity, which is unique to DNA repair, has been christened **excision nuclease** or **excinuclease** to clearly distinguish it from other endonucleases.

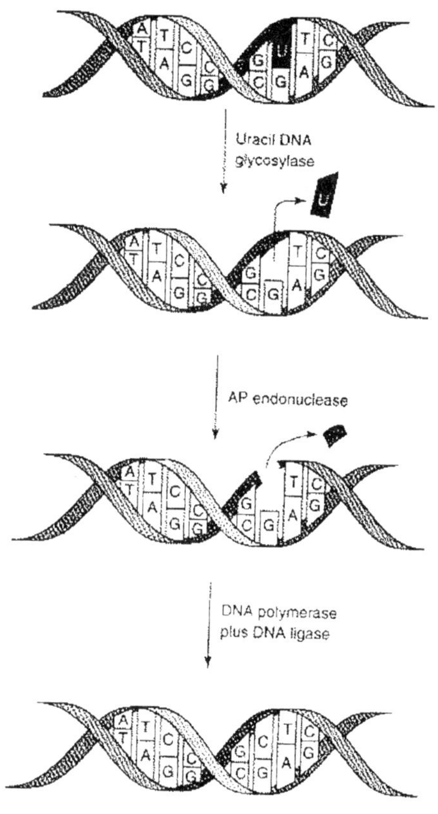

Uracil DNA glycosylase repair of DNA.

Now let us study about mismatch repair in prokaryotic and eukaryotic cells.

Mismatch repair in both prokaryotic and eukaryotic cells deals with errors created during DNA replication. In effect, three serially operating mechanisms—base selection, exonucleolytic proofreading, and postreplicative mismatch repair-participate in ensuring fidelity of replication. The mismathc repair system recognizes and eliminates mispairing from newly synthesized DNA strands, improving the fidelity of the synthesis. Base selection and proofreading act more effectively against transversion than transitions, whereas mismatch repair does the opposite. DNA replication errors are difficult to recognize because mismatches consist of erroneous but unaltered base structures. The repair system relies on other signals with the helix to identify the newly synthesized strand, which by definition harbors the replication error. Such signals are provided in *E. coli* by a methylation reaction catalyzed by **Dam methylase** that modifies GATC sequences by introducing a methyl group at the *N*-6 position of adenines. Shortly after replication these GATC sequences exist in an unmethylated state that betrays the newly synthesized nature of the DNA strand and permits strand discrimination by the mismatch repair system.

The mismatch repair system in *E. coli* includes several different protein components, which repair mismatches in the vicinity of a GATC sequence according to complementary rules dictated by the base sequence of the methylated (i.e., preexisting) parental strand. Proteins that catalyze the process of mismatch repair have been named **MutS, MutH**, and **MutL**. Repair is initated by binding of MutS to the mismatch followed by the addition of MutL. Formation of the MutS-MutL complex activates a latent GATC endonuclease activity, vested in the MutH protein, that nicks the unmodified strand at a hemimethylated GATC site. The strand break, which can occur on either side of the mismatch, will take place as long as the mismatched base is located within the general vicinity of the GATC site, which means within a few hundred base pairs from the GATC sequence. This nick marks the strand that will be excised. When the mismatch is located on the 5′ side of the cleavage site the unmethylated strand is unwound, degraded, and replaced by new DNA synthesized in the 3′→ 5′

direction until the mismatch is reached and excised. This reaction requires a DNA helicase II, referred to also as the **MutU** protein, a 3′→5′ exonuclease (exonuclease I), DNA polymerase III, and finally DNA ligase to seal the repaired strand. If the mismatch is located on the 3′ site of the cleavage, a series of completely analogous steps takes place, except that a 5′→3′ exonuclease (**RecJ**) replaces exonuclease I (an exonuclease with both 5′→3′ and 3′→5′ activity, exonuclease III can also substitute for RecJ in the latter repair). This unusual bidirectional excision activity of the mismatch repair system suggests that this system "keeps track" of the side on which the mispair of the GATC sequence signal is located.

DNA is constantly being damaged. In the absence of efficient repair, this may be the cause of as much as 90% of all human cancers. The importance of defective mismatch repair in the development of certain types of human cancer has been demonstrated recently. Tumors associated with hereditary nonpolyposis colorectal cancer (HNPCC), which causes cancer predisposition and certain sporadic cancers, have been found to be prone to mutation by as much as two orders of magnitude higher than normal human cells. These high mutuation rates have been found to be consistently associated with deficiencies in mismatch repair.

The loss of mismatch repair fidelity is a central step in the development of HNPCC tumors has been concluded from the finding that the majority of these tumors are attributable to defects at any one of four different human genome loci. These are the *hMSH2* gene, which codes for a protein homolog of bacterial MutS protein, and the *hMLH1, hPMS1, and hPMS2* genes, which specify three similar but distinct MutL analogs. These findings demonstrate that the primary event in the development of HNPCC tumors is the loss of critical mismatch repair activity. Inefficiencies in DNA repair presumably lead to mutations that circumvent the regulatory systems controlling cell proliferation. The link between mismatch repair and the development of colon cancer provides support for the hypothesis that cancers are initiated when cells accumulate a certain mutation load. A current emphasis in studies of cancer is the search for and study of particular genes, the mutations of which appears to lead to cancer. The new findings, which demonstrate the importance of mismatch repair defects in the development of cancers, may now expand the search from simply attempting to decipher the role of certain genes in carcinogenesis to also asking why and how some cells accumulate an excessive number of mutations.

9.11 Replication of DNA

The double stranded structure of DNA permits each strand to serve as a template for the synthesis of a new strand identical to the other strand. The correctness of this overall scheme of replication has solidly been established.

Three possibilities by which information transfer could take place during replication were initially visualized. Conservative replication could, in principle, yield a product consisting of a double helix of the original two strands and a daughter DNA

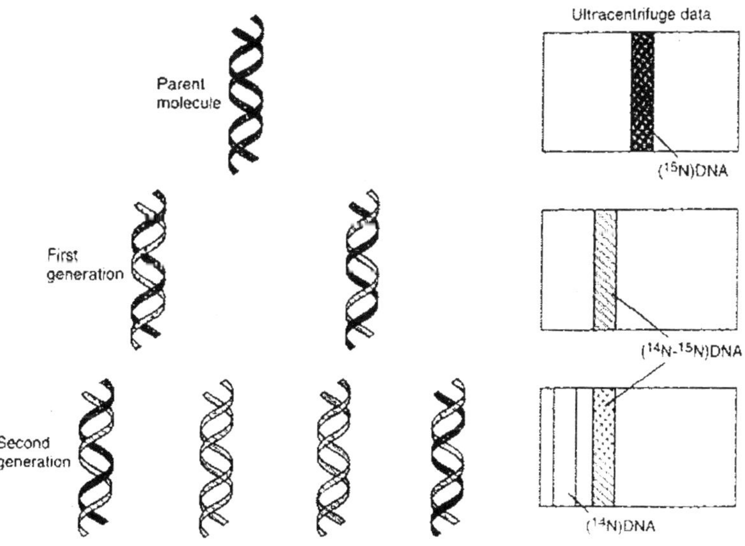

Semiconservative replication of DNA.

consisting of completely newly synthesized chains. A second possibility, labeled dispersive, would have resulted if the nucleotides of the parental DNA were randomly scattered along the strands of the newly synthesized DNA. The synthesis of DNA eventually proved to be a **semiconservative process.** After each round of replication; the structure of parental DNA is found to preserve one of its own original strands combined with a newly synthesized complementary polynucleotide.

The semiconservative nature of replication was elegantly suggested by a classic experiment that allowed the physical separation and identification of the parental and the newly synthesized strands. *Escherichia coli* was grown in a medium containing [^{15}N]-ammonium chloride as the exclusive source of nitrogen. Several cell divisions were allowed to occur, during which the naturally occurring ^{14}N in the DNA of *E. coli* was, for all practical purposes, replaced by the heavier ^{15}N isotope. The ^{14}N-containing nutrient was then added, and cells were removed at appropriate

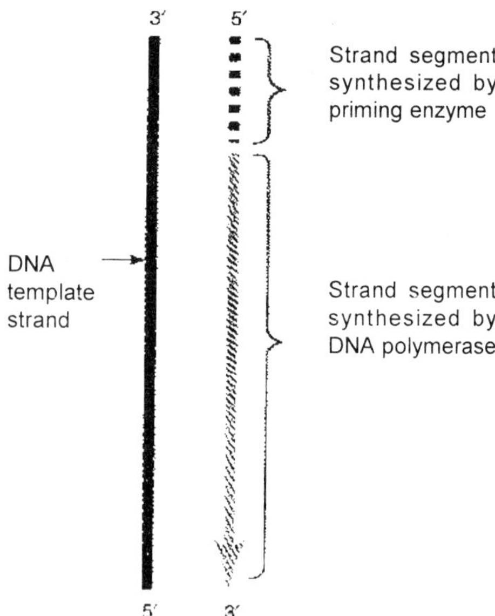

Synthesis of primer for DNA replication.

intervals. The DNA of these cells was extracted, and the ratios of ^{14}N to ^{15}N content were determined by equilibrium density gradient centrifugation.

In some bacterial systems and phages, the priming enzyme has activity characteristic of an RNA polymerase because the ribonucleotides condense to form the primer. In other systems the primase does not discriminate between 5′-ribonucleotides and 5′-deoxyribonucleotides. As a general rule, however, primases use ribonucleotides for incorporation into primers. Some enzymes that catalyze the synthesis of primers act exclusively as primases, while other possess additional enzymatic activities. In mammalian cells primase activity is vested in **DNA polymerase** α, an enzyme that is also involved in DNA stand elongation and in DNA repair. Once the primers have been synthesized, the DNA polymerase can move in and take over the process of synthesis. It is not clear what signal causes a switchover from primase to DNA polymerase, although it has been suggested that a specialized ribonuclease (**RNaseH**) is involved.

If DNA polymerase were the enzyme that would begin DNA synthesis by laying down the very first nucleotide complementary to the template, the efficiency of DNA synthesis would be severely reduced. Since the bases in a very short segment of a double helix have high configurational flexibility, the first nucleotide introduced into a newly synthesized DNA strand would likely be mispaired and would immediately activate the proofreading activity of DNA polymerase. The outcome would be a fruitless back-and-forth cycle of synthesis and proofreading by DNA polymerase with little net synthesis of new DNA. In contrast, primases, which have no proofreading ability, can quickly and efficiently position primers that can be elongated with DNA polymerases without appreciable backtracking. The primases ignore mismatches and produce an RNA chain long enough to allow the DNA polymerase to operate at the 3′ end of a double-stranded structure that restricts newly introduced nucleotides on the basis of strict complementary rules. The mismatches introduced by the primase are irrelevant because the characteristic RNA-like structure of primers allows for their subsequent wholesale removal and replacement by DNA of an equivalent composition.

Although primers are almost invariably short RNA or RNA-like segments, RNA priming is not used universally. In the "rolling circle" replication mechanism of DNA, a 3′-OH primer is generated by endonuclease digestion of parental DNA, and with **parvoviruses** a 3′-OH primer is generated by the folding back of an existing 3′–terminus. A single deoxyribonucleotide can serve as primer in **adenovirus.** Such a nucleotide with its 3′-OH terminus free, is attached to the end of a template strand through a virus-encoded specific protein.

Extensive studies in *E. coli* and its phages have permitted the proposal of a replication model that depends on the action of a large number of proteins. With the specific exceptions noted in the sections that follow, this model may also be viewed as a basic scheme for DNA replication is most other cells.

Synthesis of DNA begins at a specific site of the chromosome referred to as the **replication origin,** which in *E. coli* is referred to as **OriC**. Initiation of DNA synthesis involves participation of as many as 20-30 different proteins, many of which are needed to be present at the origin of replication in multiple copies. OriC must be recognized by specific proteins, and the origin must unwind to allow helicase, primase, and DNA polymerase III to have access to each DNA strand OriC is a sequence of 245 base pairs that contains four sites (nucleotide 9-mers with a similar nucleotide sequence) at which **dnaA,** a tetramer consisting of four identical subunits, can initiate the stepwise assembly of all the proteins and enzymes necessary to carry out replication. In addition, the origin contains 11 methylation sites recognized by **Dam methylase** and three AT-rich direct tandem repeats consisting of 13 base pairs each. This final assembly is called a **replisome.**

Formation of a replisome begins with the binding of one dnaA molecule at each one of the 9-mers, provided that these binding sites are fully methylated. The dnaA apparently recognizes these 9-mers on the basis of their conformation, which appears to be slightly curved with the double helix somewhat elongated relative to typical B-DNA. Several more additional dnaA molecules are then added via a highly cooperative process to form a nucleosome-like structure. An additional factor, **HU protein,** participates in the formation of this complex.

The dnaA and HU protein interact with the OriC in a manner that promotes the opening of the DNA strands in the AT-rich regions adjacent to the origin. Finally, dnaA, with the aid of **dnaC**, adds **dnaB** in the complex. The dnaB, by virtue of its helicase activity, creates an initiation "bubble" consisting of a few hundred nucleotide pairs. The energy for the formation of the "bubble" is provided by ATP in a reaction catalyzed by topoisomerase II, and the 'bubble" is stabilized by SSB proteins.

Synthesis of an RNA primer begins with the formation of a prepriming complex. The prepriming assembly consists of the dnaB-dnaC complex to which four other proteins (polypeptides n, n′, n′′, and i) have been added. Addition of primase, **dnaG,** converts the prepriming complex to a **primosome.** The primosome is formed by binding of primase, together with a complex of dnaB and dnaC proteins, at specific sequences of DNA that serve as sites for formation of RNA primers. Additional factors, described as n proteins, are specific primosomal components that are responsible for placing the primosome at the appropriate sequences. In effect, the primosome "searches" the DNA for these sequences at the expense of ATP. Once the correct destination of the primosome is reached, RNA primer synthesis is initiated.

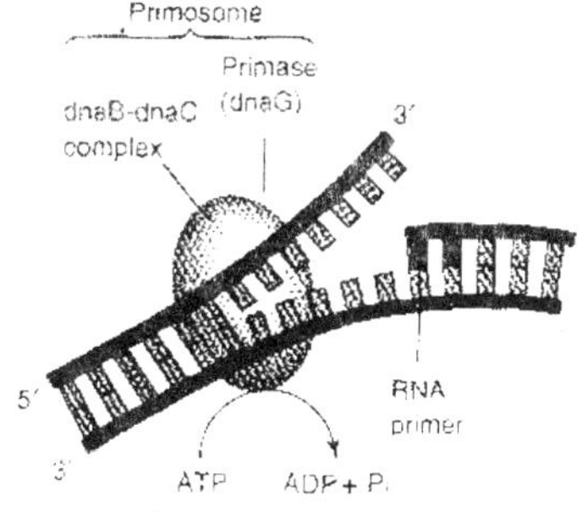

Primosome of *E. coli.*

The primosome interacts with a template, at each one of the two forks generated by the formation of a "bubble" and begins the synthesis of RNA primers on the two leading strands. Assembly of the replisome is completed by addition to the primosome of DNA polymerase III and **rep proteins.**

Initiation can be regulated by either restricting the availability of dnaA-binding sites at OriC or by limiting the concentration of dnaA. **Methylation** provides a switch for the availability of dnaA-binding sites. Once replication has been initiated, the dnaA near OriC binds to the plasma membrane and becomes unavailable to Dam methylase. In addition, binding of DNA in the vicinity of OriC to the cellular membrane sequesters the *dnaA* gene, which is situated near OriC. As a result, the synthesis of dnaA protein is inhibited and its cellular concentration is lowered.

Initiation of the leading DNA strand at OriC by the primosome is more complex than the subsequent initiation of synthesis of Okazaki fragments on the lagging strand initiated by primase at sites selected by the prepriming proteins. The initiation of the leading strand does not present the cell with serious topological problems, but for continuation of synthesis helicase II and rep protein are essential. These enzymes unwind and separate the strands in each of the two forks created by the initiation event. As the helicases move in advance of each fork, two single-stranded regions are generated on parental DNA. These regions are immediately covered by single-strand binding protein that keeps the fork open and allows DNA polymerase III to take over the elongation of primers. A signal for initiation of the lagging strand, uncovered on the template by the movement of helicase, leads to the binding of primase. Primase, the action of which is triggered by the prepriming proteins, synthesizes a brief complementary segment of the strand. This segment serves as a primer for covalent extension of the strand synthesized by DNA polymerase III and for formation of Okazaki fragments. DNA polymerase III complexes are endowed with similar but somewhat distinct properties, one tailored for the continuous synthesis of the leading strand and the other for the discontinuous synthesis of the lagging strand. This polymerase assembly, which appears to combine primase activity with nonidentical twin active sites for polynucleotide synthesis, allows for concurrent replication on both strands. In this scheme, looping of the lagging strand template by 180° brings it to the same orientation as the leading strand template. Thus a primer synthesized at the lagging strand is drawn past it. When a nascent (Okazaki) fragment reaches the 5′ end of the previously synthesized Okazaki fragment, the lagging strand template is released and unlooped. Removal of the primer portions at the 5′ end of the Okazaki fragments by DNA polymerase I, repair by the same enzyme and joining of the repaired fragments by DNA ligase produces intact DNA strands. The DNA synthesis in eukaryotes appears to be a process that is fundamentally similar to that occurring in prokaryotes. Formation of a replication fork, primer synthesis, Okazaki fragments, primer removal, and gap bridging between newly synthesized DNA segments, all parallel the corresponding steps that occur in prokaryotes, but the overall process is quite a bit more complex. Replication among eukaryotes, from yeasts to humans, shares similarities.

As expected, differences are more pronounced between prokaryotes and eukaryotes. In rapidly growing prokaryotes, DNA is replicated through much of the cell cycle and cell division occurs as soon as DNA synthesis has ceased. In contrast, eukaryotic DNA synthesis (and histone synthesis) is confined to only one part of the cell cycle, specifically the synthetic (S) phase of the interphase. This phase is preceded and followed by two periods during which DNA is not synthesized (gap periods G1 and G2). Cell division occurs at a different time within the interphase, referred to as the mitotic (M) period. Beyond this characteristic limitation of eukaryotic replication to a certain period of the cell cycle, important differences in replication between prokaryotes and eukaryotes arise primarily from the larger size of eukaryotic DNA (about 10^5–10^6 kb content) as compared to prokaryotic DNA (about 5×10^3 kb for *E. coli*), the distinct packaging of eukaryotic DNA in the form of chromatin, and the slower rates of fork movement in eukaryotes. For DNA to become available to DNA polymerases, nucleosomes must disassemble, a step that slows the rates of fork movement.

DNA polymerase movement does not exceed 30,000 base pairs per minute, which is considerably slower than the rates observed for *E. coli.* Based on the higher DNA content of animal cells, and the lower activities of DNA polymerases in comparison to bacteria, the replication cycle of eukaryotic cells could be expected to take as long as a month to complete. In fact, however, the replication cycle is completed within hours, because compensating factors are in operation. Eukaryotic cells contain a large number of DNA polymerase molecules (often in excess of 20,000) as compared to a few dozen in each *E. coli* cell. DNA polymerase initiates bidirectional synthesis but at several origins of replication located anywhere between 5 and 300 kilobase pairs (kb) apart within the chromosome, depending on species and cell type. DNA segments between two origins of replication are termed **replicons.** An average human chromosome contains as many as 100 replicons and replication may proceed simultaneously at as many as 200 forks.

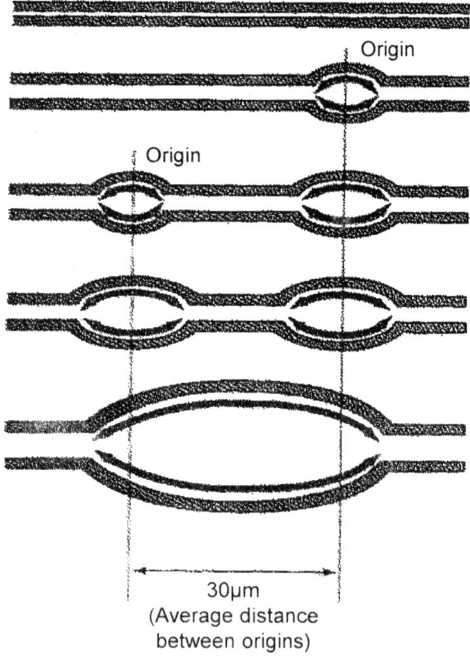

Replication of mammalian DNA.

More origins can be found in developmentally active cells that carry out DNA synthesis at very rapid rates. During early embryogenesis the largest chromosome of *Drosophila melanogaster* contains as many as 6000 replicating forks, or one for every 10 kb.

DNA synthesis directed by circular mtDNA, and in some instances by bacteria and viruses, gives rise to linear daughter DNA molecules that contain the base sequence of parental DNA repeated numerous times. These repeated linear DNAs, which are known as **con-catemers,** are essential for the bacterial mating and may be involved in gene amplification. The synthesis of concatemer DNA occurs by a mechanism known as **rolling circle replication.**

Synthesis of the lagging strand is catalyzed by **DNA polymerase α**. This polymerase has similar structure and properties in all eukaryotes. The large subunit (mol wt ~180,000) of the tetrameric DNA polymerase α is vested with the usual 5′ → 3′ nucleotide polymerizing activity. Polymerase *α*, isolated from some but not most sources, also has a 3′ → 5′ exonuclease activity. Two of the other subunits of the enzyme are **primases.** The primary proofreading function in eukaryotes appears to be carried out by polymerase δ. Polymerase δ improves the fidelity of replication by a factor of 10^2 and contributes in limiting the rates of overall error to 10^{-9} to 10^{-12}.

The relatively low processivity of DNA polymerase α is typical for an enzyme involved in synthesis of the lagging strand that is assembled from segments of DNA that are no larger than 100-200 bp. The size of these Okazaki fragments is approximately equal to the length of DNA wrapped around a nucleosome. This observation suggests that eukaryotic DNA may be releasing one nucleosome at a time for priming of the lagging chain. The primase subunit of the enzyme synthesizes Okazaki segments as a closely coordinated priming-synthesizing activity, by laying down RNA primers containing 5-15 nucleotides that are subsequently extended by the synthetic activity of polymerase α. This polymerase catalyzes the synthesis of a polynucleotide chain at a rate of 50 nucleotides per second, which is about 1/20 the rate of *E. coli* DNA polymerase III synthesis. Looping of the lagging strand allows a combined polymerase α–polymerase δ asymmetric dimer to assemble and elongate both the leading and lagging strands in the same overall direction that corresponds to the direction of the fork movement.

9.12 Recombination of DNA

Recombination of DNA is a process during which genetic material is rearranged by breaking and joining portions of the same DNA molecule or portions of different DNA molecules.

Recombination takes place between the DNAs of different organisms to generate a new composite DNA.

Homologous genetic recombination produces an exchange between a pair of distinct DNA molecules, often two slightly variant copies of the same chromosome, or two segments of DNA generated from the same DNA molecule. The main requirement for this process to occur is that the recombining DNAs are **homologous.** This means that the two DNAs share very similar base sequences over an extended region that may contain several thousand bases. An important example of homologous recombination in eukaryotes is the exchange of sections of homologous chromosomes during the early development of gametes (egg and sperm cells). In this manner slightly different versions of the same gene (alleles) can evolve during meiosis. Gene "mixing and reassortment" by general recombination is also widespread in bacteria. Homologous recombination is quite complex and involves a multistep mechanism catalyzed by a large number of different proteins. Prominent among them is the RecA protein, which also participates in SOS DNA repair.

Conservative site-specific recombination or **site-specific recombination** requires the presence of only short homologous DNA sequeces. However, site-specific recombinations occur only in specific DNA sequences present in both participating DNA molecules. The process is catalyzed by enzymes known as **recombinases.**

Transpositional site-specific recombination, or simply **transposition,** differs from conservative site-specific recombination in that it does not require a specific DNA sequence in the "target" chromosome. Transposition is catalyzed by **transposases.** Both transposases and recombinases recognize and act on specific DNA sequences. Recombination of either type is responsible for the insertion of viruses, plasmids, and **transposable elements (transposons)** into chromosomal DNA. Transposons are DNA elements that can move from location to location within a genome, in both bacteria and eukaryotes. Viruses are related to plasmids and transposons but also differ from these genetic elements in that viruses can synthesize a protein coat that allows them more host-independent existence. Plasmids and transposons are confined to replicate only within a specific cell and the progeny of that cell.

The most common recombination is the **homologous** type, **Site-specific recombination** and **transposition** are relatively rare, but important, events in that they may control replicative function in some viruses and certain aspects of development. Homologous recombination generates new combinations of genes that can lead to genetic diversity. DNA mutation and recombination are Homologous recombination, which is accompanied by the formation of a **heteroduplex DNA** region, clearly requires breaking and rejoining of chromosomal DNA. Recombination occurs via a fairly complex multistep mechanism. This scheme gives a minimal overview of recombination, in that each of the steps shown may represent more than one enzymatically catalyzed process. Numerous gene products are involved in homologous recombination.

Recombination may begin by introduction of a single-strand nick at a selected site of one of the DNA duplexes undergoing recombination. The resulting 3′-ended single-strand tail can then invade a homologous DNA duplex. Homologous DNA duplexes are chromosomes with the same linear arrangement of genes but with base sequences that may differ between the two duplexes. The variance is usually minor and may consist of no more than one different base among the millions of base pairs present in the chromosome. Single strand invasion places the homologous DNA duplexes side by side in a process referred to as **synapsis.** Synapsis does not necessarily involve contacts between homologous sequences and further movement of the DNAs with respect to each other may be necessary until homologous sequences come into contact. This process is referred to as **homologous alignment.** **Strand invasion** is accompanied by **strand displacement** in the homologous DNA duplex resulting in the formation of a so-called **D-loop.** The "D-loop" strand that has been displaced by strand invasion is now nicked and it pairs with its complementary strand in the original duplex. The ends of exchanged strands are then ligated to form a stable **cross-stranded intermediate** known as **Holliday junction.**

The junction can migrate in either direction by unwinding and rewinding of the two duplexes to produce a further exchange of single strands between interacting chromosomes. This process is known as branch migration, results in strand exchange and it produces heteroduplex regions of varying lengths.

The resulting heteroduplex, can also be presented in another form that is generated by merely pulling the ends of the heteroduplex together. A twist of this structure produces an isomeric heteroduplex, which is called the **Chi** form. In order to resolve the Chi form two additional single-strand nicks can be made, in either the horizontal direction or vertical direction, leading to two distinct products. Gaps present in these structures are repaired and ligated, leading to either one of the two products. The manner in which nicks are introduced in the horizontal and vertical directions is fundamentally different. In one case (horizontal direction) nicks are introduced again into the strands that were initially nicked, although at different sites, producing two duplexes in which one strand of each remains intact. These duplexes contain heteroduplex regions, generated by branch migration, that are misleadingly referred to as **Patch recombinant heteroduplexes.** These duplexes contain the same genes and in the same linear order as the initial duplexes. In vertical direction nicks, the complementary strands that previously were left intact are nicked again (though at different sites), producing two duplexes of true recombinant DNA, referred to as **splice recombinant heteroduplexes.** In these true recombinant heteroduplexes are linear order of DNA sequences contained in the original duplexes is clearly rearranged.

Support for this multistep recombination scheme has accumulated over the years based on genetic investigations, on electron microscopy of Holliday junctions, and by isolation of proteins and enzymes that can catalyze many of the transformations described in this recombination scheme.

Homologous recombination in *E. coli* requires about 25 enzymes for recombination. A partial list includes **RecA** protein, **RecBCD** enzyme (which is the product of three distinct *E. coli* genes, *recB, recC,* and *recD*), **RuvAB** and **RuvC** proteins, DNA polymerase I, DNA gyrase, DNA topoisomerase I, DNA ligase, and DNA helicases. Proteins homologous to RecA have also been isolated from yeast and human cells.

Homologous recombination in *E. coli* begins with RecBCD, which is a site-specific **endonulease** and an ATP-dependent **helicase**. RecBCD can initiate recombination by unwinding DNA and, on occasion, cleaving one strand. The enzyme binds to one end of linear DNA and travels along the helix at the expense of ATP, unwinding DNA as it moves and rewinding DNA behind it at a slower rate than unwinding. This produces a "bubble" consisting of two single-stranded loops that propagate on the DNA with the advance of the RecBCD. *Escherichia coli* DNA is characterized by the presence of about 1000 copies of the sequence 5′-GCTGGTGG-3′ that, on average, occurs at intervals of 4-5 kb. These Chi sites are **"hot spots" for recombination** as they increase the frequency of recombination. When the advancing RecBCD encounters a Chi site within a "bubble," it cleaves the DNA strand that incorporates the 5′-GGTGGTGG-3′ sequences 5-6 nucleotide to the 3′ side of the Chi site. The helicase activity generates a 3′ single-stranded tail of DNA that is progressively lengthened to several kilobases.

This growing single-stranded tail can then initiate the **strand invasion** process with the assistance of RecA, which catalyzes a multiplicity of reactions in DNA recombiantion. RecA interacts with single-stranded (ss) and double-stranded (ds) DNA and catalyzes **pairing of homologous DNA sequences, invasion of ssDNA** into the homologous double helix, formation of the **Holliday junction,** and migration of this junction (branch migration). These activities of RecA depend on the presence of a RecA site that recognizes ssDNA and promotes the cooperative binding of the protein to ssDNA. Formation of a long and relatively still **nucleofilament** prevents the ssDNA tail from reassociating with the complementary strand within the DNA duplex, from which it originated, and prepares the single strand for invasion. In the resulting nucleofilament that binds one RecA molecule per 3 bases, the polynucleotide is positioned within a deep groove of the RecA protein. A second site on RecA recognizes and binds preferentially to dsDNA. In this nucleofilament each RecA monomer covers six nucleotides and each successive monomer binds to the opposite site of the DNA helix. For the sake of simplicity the dsDNA is shown as free from RecA. The RecA-ssDNA and RecA-dsDNA nucleofilaments differ in their geometry from B-DNA, but both filaments represent partially unwound and unstacked helical structures that are extended lengthwise by 50% relative to B-DNA. DNA unwinding in the RecA-dsDNA nucleofilament (to about 18.6 bp per turn) exposes H-bond tonors and acceptors in the major groove of the double helix, making them available for interaction with the ssDNA-RecA filament. Thus RecA contributes to the recognition of regions of homology between DNA strands. Once **homologous alignment** is established, a fairly stable **triple-stranded intermediate** can be formed. In this structure the third strand is in contact with the major groove of the duplex, aligned in a manner that permits RecA to flip the base pairing of the two identical strands.

The flipping of the base pairing and the resulting invasion of the RecA-ssDNA filament involve the exchange of two identical (or nearly identical) strands between helical structures, which therefore requires an ordered rotation of two aligned strands. The polynucleotides are prepared for the exchange by "the extended" conformation generated by RecA. Strand exchange can be extended by **branch migration,** which means that progression of the exchange requires both invasion and branch migration. Branch migration may be described as a process in which an unpaired region of a single DNA strand displaces a DNA strand from a region of homologous dsDNA and moves the branching point, without appreciably increasing the total number of disrupted base pairs. Migration is achieved by RecA-catalyzed rotation of RecA-bound DNA strands involved in the exchange. The resulting "spooling" action, in which topoisomerases may be involved, moves the branch as ATP is hydrolyzed.

Branch migration also occurs at the Holiday junction that is subsequently formed. In this intermediate homologous DNA helices that were initially paired are held together by mutual exchange of two of the four strands. Stereochemistry of the intermediate is determined by the juxtaposition of the grooves and the phosphate backbones of the participating helices, and the point of exchange or actual junction can be moved back and forth along the helices. Migration of the junction can proceed in the absence of RecA. This RecA-independent migration of the junction is catalyzed by a complex of **RuvA** and **RuvB.** RuvA binds to the junction and acts as a specificity factor that targets RuvB, which is an ATPase, to the junction. The RuvAB complex promotes migration and increases the length of the heteroduplex DNA at the expense of ATP. Finally, the Holliday junction is recognized and resolved into products by the **RuvC endonuclease,** a dimer of 19-kDa subunits related to each other by a dyad axis of symmetry. The catalytic center of this **resolvase** lies at the bottom of a cleft that fits a DNA duplex. Only strands with the same polarity are cleaved and produce two types of heteroduplex molecules, one type in which only single-strand segments are exchanged **(patch recombinants)** and another type, a true recombinant, in which the ends of molecules have been exchanged **(splice recombinants)**. Resolution is completed by DNA polymerase I, DNA topoisomerase I, DNA gyrase, and DNA ligase.

RecA also exhibits a highly specific protease activity that is activated by unpaired DNA strands and is directed at specific regulatory proteins.

10

General Bio-chemical Mechanism

10.1 Metabolism and Function of Nervous Tissue

Knowledge of the chemical composition of the **brain** began with the work of J.L.W. Thudichum in 1884 and the publication of his monogram. "A Treatise on the Chemical Composition of the Brain, Based Throughout on Original Research" (cited in West and Todd, *Textbook of Biochemistry,* MacMillan, 1957). Thudichum's research was supplemented with the work of others during those earlier years. There have been almost explosive advances during more recent years, through the use of molecular biological techniques, not only in our knowledge of the composition of the brain but also of molecular mechanisms involved in many brain/neuronal functions.

About 2.4% of an individual's body weight is nervous tissue, of which approximately 83% is the brain. The **nervous system** provides the communications network between the senses, the environment, and all parts of the body. The brain is the command center. This system is always functioning and requires a large amount of energy to keep it operational. Under normal conditions, the brain derives its **energy** from **glucose metabolism. Ketone bodies** can cross the **blood-brain barrier** and be metabolized by brain tissue. Their metabolism becomes more prominent during **starvation,** but even then they cannot replace the need for glucose. The human brain uses approximately 103 g of glucose per day. For a 1.4-kg brain, this corresponds to a rate of utilization of approximately 0.3 μmol min^{-1} g^{-1} of tissue. This rate of glucose utilization represents a capacity for **ATP production** through the

tricarboxylic acid (TCA) cycle alone of approximately 6.8 μmol min^{-1} g^{-1} of tissue. Of course, the TCA cycle is not 100% efficient for ATP production, nor is all of the glucose metabolized through it. Most of the ATP used by the brain and other nervous tissue is generated aerobically through the TCA cycle, which functions at near maximum capacity. **Glycolysis** functions at approximately 20% capacity. Much of the energy used by the brain is to maintain ionic gradients across the palsma membranes, to effect various storage and transport processes, and for the synthesis of neurotransmitters and other cellular components.

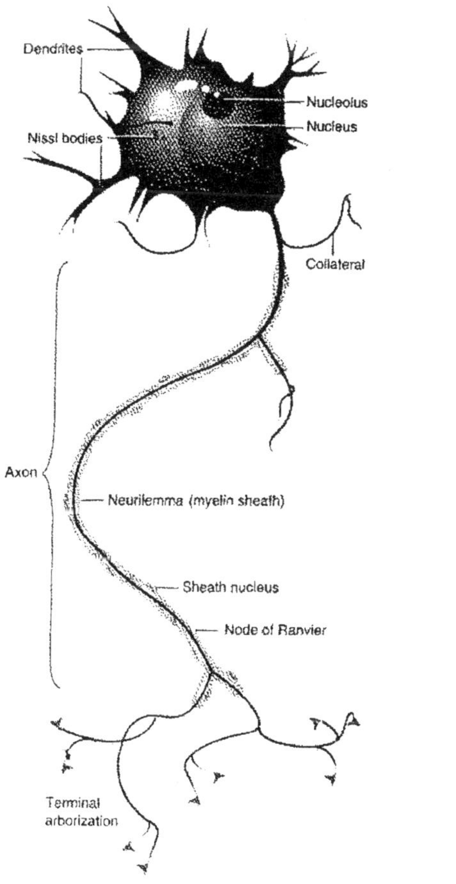

A motor nerve cell and investing membranes.

Two features of brain composition are worth noting. It contains specialized and **complex lipids,** but they appear to function to maintain membrane integrity rather than to have metabolic roles. There is generally a rapid **turnover rate** of **brain proteins** relative to other body proteins in spite of the fact that the cells do not divide after they have differentiated.

Cells of the nervous system responsible for collecting and transmitting messages are the **neurons.** They are very highly specialized. Each neuron consists of a cell body, **dendrites** that are short antenna-like protrusions that receive signals from other cells, and an **axon** that extends from the cell body and transmits signals to other cells. The central nervous system (CNS) is a highly integrated system where individual neurons can receive signals from a variety of different sources, including both inhibitory and excitatory stimuli.

Cells other than neurons exist in the CNS. In the brain, there are about 10 times more glial cells than there are neurons. Glial cells occupy spaces between neurons and provide some electrical insulation. Glial cells are generally not electrically active, and they are capable of division. There are basically five types of glial cells: Schwann cells, oligodendrocytes, microglia, ependymal cells, and astrocytes. Each type of glial cell has a specialized function, but only astrocytes appear to be directly associated with biochemical functions related to neuronal activity. One is metabolic and the other anatomical.

Astrocytes send out processes at the external surfaces of the CNS. These processes are linked to form anatomical complexes that provide sealed barriers and isolate the CNS from the external environment. Astrocytes also send out similar processes to the circulatory system, inducing the endothelical cells of the capillaries to become sealed by forming tight junctions that prevent the passive entry into the brain of water-soluble molecules. These tight junctions form what is commonly known as the **blood-brain barrier.** Water-soluble compounds enter the brain only if there are specific membrane transport systems for them.

10.2 Metabolism and Vision

The eye, our window to the outside world, allows us to view the beauties of nature, the beautifies of life, and *vide* this textbook, the beauties of biochemistry. What are the features of this organ that permit this view? A view through any window, through any camera lens, is clearest when unobstructed. The eye has evolved in such a way that a similar objective has been achieved. It is composed of live tissues that require continuous nourishment for survival. Energy and metabolites for growth and maintenance are derived from nutrients by conventional biochemical mechanisms, but the structures responsible for these processes are arranged and distributed such that they do not interfere with the visual process. Also, the brain has devised an enormously efficient filtering system that makes invisible objects within the eye that may appear to lead to visual distortion. In addition, different tissues use specific metabolic pathways to accommodate their unique needs.

Light entering the eye passes progressively through the **cornea;** the anterior chamber, which consists of the **aqueous humor;** the lens; the vitreous body which consists of the **vitreous humor;** and finally focuses on the **retina,** which contains the visual sensing apparatus. The exterior of the cornea is bathed by tears, while the interior is bathed by the aqueous humor, an iso-osmotic fluid containing salts, albumin, globulin, glucose, and other constituents. The aqueous humor brings nutrients to the cornea and to the lens,

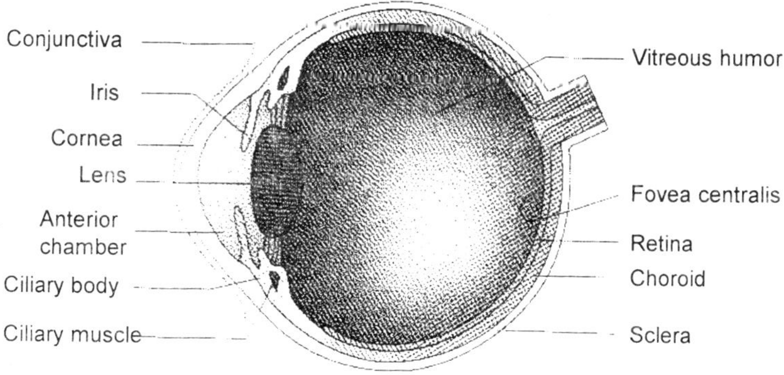

Schematic of a horizontal section of the left eye.

and it removes end products of metabolism from them. The vitreous humor is a collagenous or gelatinous mass that helps maintain the shape of the eye while allowing it to remain somewhat pliable.

The eye is an extension of the nervous system, and like other tissues of the central nervous system, the major metabolic fuel is glucose. The cornea, which is not a homogeneous tissue, obtains a relatively large percentage of its ATP from aerobic metabolism. About 30% of glucose used by the cornea is metabolized by glycolysis and about 65% by the **hexose monophosphate pathway.** On a relative weight basis, the cornea has the highest activity of the hexose monophosphate pathway of any other mammalian tissue. It also has a high activity of **glutathione reductase,** an activity that requires NADPH, a product of the hexose monophosphate pathway. Corneal epithelium is permeable to atmospheric oxygen, that is necessary for various oxidative reactions. The reactions of oxyen can result in the formation of various **active oxygen species** that are harmful to the tissues, perhaps in some cases by oxidizing protein sulfhydryl groups to disulfides. Reduced glutathione (GSH) is used to reduce those disulfide bonds back to their original native states while GSH itself is converted to oxidized glutathione (GSSG). Furthermore, oxidized glutathione (GSSG) may also be formed by auto-oxidation. Glutathione reductase uses NADPH to reduce GSSG to 2GSH.

The activities of the hexose monophosphate pathway and the glutathione reductase maintain this tissue in an appropriately reduced state by effectively neutralizing the active oxygen species.

The **lens** is bathed on one side by the aqueous humor and supported on the other side by the vitreous humor. The lens has no blood supply, but it is metabolically active. It gets nutrients from the aqueous humor and eliminates waste into the aqueous humor. The lens is mostly water and proteins. The majority of the proteins are the α-, β-, **and** γ-**crystallins.**

The eye may be compared with a video camera. The camera collects images, converts them into electrical pulses, records them on magnetic tape, and allows their visualization by decoding the taped information. The eye focuses on an image by projecting that image onto the retina. A series of events begins, the first of which is photochemical, followed by biochemical events that amplify the signal, and finally electrical impulses are sent to the brain where the image is reconstructed in "the mind's eye." During this process, the initial event has been transformed from a physical event to a chemical reaction, through a series of biochemical reactions, to an electrical event, to a conscious acknowledgement of the presence of an object in the environment outside the body.

When photons of light enter the eye and are absorbed by photoreceptors in the **outer segments** of **rods** or **cones,** they cause isomerization of the visual pigment, **retinal,** from the 11-*cis* form to the all-*trans* form. This isomerization causes a conformation change in the protein moiety of the complex and affects the resting membrane potential of the cell, resulting in an electrical signal being transmitted by way of optic nerve to the brain.

The **photoreceptor cells** of the eye are the **rods** and the cones. Each type has flattened disks that contain a photoreceptor pigment. This pigment is **rhodopsin** in the rod cells, and **red, green,** or **blue pigment** in the cone cells. Rhodopsin is a transmembrane protein to which is bound a prosthetic group, **11-*cis*-retinal.** Rhodopsin minus its prosthetic group is **opsin.** The three proteins that form the red, green, and blue pigments of cone cells are different from each other and from opsin.

Rhodopsin, an approximately 40-kDa protein, contains seven transmembrance α–helices. The 11-*cis*-retinal is attached through a protonated Schiff base to the ε-amino group of lysine-296 on the seventh helix. Lysine 296 lies about midway between the two faces of the membrane. A 9-Å resolution three-dimensional (3-D) model for rhodopsin, obtained by cryomicroscopy, shows that most of the helices are perpendicular to the surface of the membrane.

There are central nervous system disorders associated with the Niemann-Pick group of diseases that can become evident by ocular changes. Some of these are observed as abnormal macula with gray discoloration and granular pigmentation or granule opacities about the fovea.

Acute type I Niemann-Pick disease, lipidosis with sphingomyelinase deficiency and primary sphingomyelin storage, may show a cherry red spot in the retina in as many as 50% of patients. Macula halo syndrome applies to the crystalloid opacities seen in some patients with subacute type I disease. They form a halo approximately one-half the disk diameter at their outer edge and are scattered throughout the various layers of the retina. They do not interfere with vision.

In an 11-year-old girl who had type II disease, more extensive ocular envolvement was observed. There was sphingomyelin storage in the keratocytes of the cornea, the lens, the retinal ganglion cells, the pigmented epithelium, the corneal tract, and the fibrous astrocytes of the optic nerve.

Retinitis pigmentosa is a secondary effect of the abnormal biochemistry associated with Niemann—Pick disease.

10.3 Cone Cells

The general mechanism by which light stimulates cone cells is exactly the same as it is for rod cells. There are **three types of cone cells,** defined by the visual pigments they contain, which are either blue, green, or red. Normally, only one type of visual pigment occurs in a single cell. The blue pigment has optimum absorbance at 420 nm, the green pigment at 535 nm, and the red pigment at 565 nm. Each of these pigments has 11-*cis*-retinal as the prosthetic group, and, when activated by light, the 11-*cis*-retinal isomerizes to all-*trans*-retinal in exactly the same manner as it does in the rod cells. Colours other than those of the visual pigments are distinguished by graded stimulation of the different cones and comparative analysis by the brain. Colour vision is **trichromatic.**

The characteristic of colour discrimination by cone cells is an inherent property of the proteins of the visual pigments to which the 11-*cis*-retinal is attached. The 11-*cis*-retinal is attached to each of the proteins through a protonated Schiff base. The conjugated double-bond system of 11-*cis*-retinal influences the absorption spectrum of the pigment. When 11-*cis*-retinal is bound to different visual proteins, amino acid residues in the local areas around the protonated base and the conjugated π-bond system influence the energy level and give different absorption spectra with absorption maxima that are different for the different colour pigments.

Genes for the colour pigments have been cloned and their amino acid sequences inferred from the gene sequences. Open circles represent amino acids that are the same, and closed circles represent amino acids that are different. A string of closed circles at either end may represent an extension of the chain of one protein relative to the other. The red and green pigments show the greatest degree of homology, about 96% identity, whereas the degree of homology between different pairs of the others is between 40% and 45%.

Genes encoding the **visual pigments** have been mapped to specific chromosomes. The rhodopsin gene resides on the third chromosome, the gene encoding the blue pigment resides on the seventh chromosome, and the two genes for the red and green pigments reside on the X chromosome. Abnormal colour vision results from mutations in one or more of these genes. In spite of their great similarity, the red and green pigments are distinctly different proteins. Individuals have been identified with inherited variations that affect one but not both pigments simultaneously. In addition, there may be more than one gene for the green pigment, but it appears that only one is expressed.

The person who developed the atomic theory of chemistry. John Dalton (1766-1844), was colour blind. He thought his colour blindness was due to the vitreous humor being tinted blue, selectively absorbing longer wavelengths of light. He instructed that after his death his eyes be examined to determine whether his theory was correct. An autopsy revealed that the vitreous humor was "perfectly pellucid," normal. Using DNA analysis on his preserved eyes obtained from the British Museum, it has now been demonstrated that Dalton was missing the blue pigment. Thus, instead of having trichromatic vision, he was dichromatic with a vision type referred to as **deuteranopia.** The type of colour blindness of one who is missing the green pigment is **protanopia.**

The sensitivity and the response time of the rods are different from that of the cones. Absorption of a single photon by photoreceptors in rod cells generates a current of approximately 1-3 picoamperes (1-3 $\times$ 10^{-12} pA), whereas the same event in the cones generates a current of approximately 10 femtoamperes (10 $\times$ 10^{-15} fA), about 1/100th of the rod response. The response time of cone cells, however, is about four times faster than that of rod cells.

The chromosomal arrangement of genes for vision precludes inheritance of a single defective gene from one parent that would render recipients sightless. Genes that code for visual pigments occur on chromosomes that exist in pairs except in males where there is a single X chromosome containing the genes for red and green pigments. In females, there is a pair of X chromosomes and, therefore, colour vision abnormalities in females are rare, affecting only about 0.5% of the population. By contrast, about 8% of males have abnormal colour vision that affects red or green perception and, on rare occasions, both. For the sake of simplicity, the proteins coded for by the different genes will be referred to as pigments in spite of the fact that they become visual pigments only when they form complexes with 11-*cis*-retinal.

The gene that codes for the protein moiety of rhodopsin, the rod pigment, is located on the third chromosome. Genes that code for the three pigment proteins of cone cells are located on two different chromosomes. The gene for the blue pigment is on the seventh chromosome. The genes for the red and green pigments are tightly linked and are on the X chromosome, which normally contains one gene for the red pigment and from one to three genes for the green pigment. In a given set of cones, only one of these gene types is expressed, either the gene for the red pigment or one of the genes for the green pigment.

Genetic mutations may cause structural abnormalities in the proteins that influence the binding of retinal or the environment in which retinal resides. In addition, the gene for the protein of a specific pigment may not be expressed. If 11-*cis*-retinal does not bind or one of the proteins is not expressed. the individual will have dichromatic colour vision and be colour blind for the colour of the missing pigment. If the mutation changes the environment around the 11-*cis*-retinal, shifting the absorption spectrum of the pigment, the individual will have abnormal trichromatic color vision; that is, the degree of stimulation of one or more of the three cone pigments will be abnormal. This will result in a different integration of the signal and hence a different interpretation of colour.

10.4 The Circulation of Blood

The circulation of blood is essential for life, and the integrity of the process must be maintained. Some aspects of the importance of blood circulation in the maintenance of pH, in the transport of oxygen and nutrients to cells, and in the transport of carbon dioxide and waste products from cells are well known. This section deals primarily with a description of the system responsible for clot formation and dissolution.

Blood circulation occurs in a very specialized type of closed system in which the volume of circulating fluid is maintained fairly constant. This system is also one in which the transfer of solutes across its boundaries is a necessary function. Like any system of pipes and tubes, leaks can occur and must be repaired. The process of **blood clotting** primarily addresses the question of stopping the leaks. Secondarily, small clots may form due to disease and other abnormalities that are independent of total rupture of vesicles. Discussion of the function of the process must therefore extend beyond the primary one of leak prevention to include **clot dissolution.**

The purpose of this section is to give a general picture of the mechanism of blood clotting from a biochemical viewpoint. To this end, this section will focus on the relationship between blood clot formation, blood clot dissolution, and the enzymes and other proteins involved—their activation, regulation, inhibition, and synthesis. Blood clotting is not a process of signal transduction in the same sense as are the other topics of this chapter. Instead, it is a dynamic process of signal amplification and modulation. Some of the primary questions to be addressed are: (1) What initiates the clotting process? (2) What substances, reactions, and mechanisms are responsible for forming the clot? (3) What factors and mechanisms are involved in inhibiting the clotting process once it is initiated? (4) How is the clot dissolved?

It is important for the body to maintain **hemostasis,** that is, no bleeding. Thus the process of blood clotting is designed to stop as rapidly as possible the loss of blood following vascular injury. When such an injury occurs, three major events take place: (1) **aggregation** of a protein, **fibrin,** into an insoluble network, or clot, to cover the ruptured area to prevent the loss of blood; (2) **clumping** of **blood platelets** at the site of injury in an effort to form a physical plug to stop the leak; and (3) **vasoconstriction** in an effort to reduce the blood flow through the area. Equally important is regulation of the process to prevent excessive clot formation.

The processes mentioned above are emergency mechanisms for stopping the loss of blood. The process is not complete, however, until the ruptured vessel itself is repaired and the clot dissolved. Many of the proteins involved in blood coagulation contain **epidermal growth factor (EGF)-like domains.** Whether these EGF-like domains act directly to facilitate the regrowth of blood vesicles is not clear.

Some of the major proteins (players) involved in this process (silent drama) not necessarily in order of appearance. All are important and, as time goes on, others are sure to be added. In fact, protein Z that occurs to a larger extent in children could be added but its role and function are not clear.

Clot formation initially follows two separate pathways **intrinsic** or **contact factor pathway** and **extrinsic** or **tissue factor pathway.** These pathways merge with the formation of factor Xa, the proteinase component of the multienzyme complex that catalyzes the formation of thrombin from prothrombin. From this point on, there is a single pathway for clot formation. Historically, the term intrinsic pathway came from the observation that blood clotting would occur spontaneously when blood was placed in clean glass test tubes, leading to the idea that all components for the clothing process were intrinsic to the circulating blood. Glass contains **anionic surfaces** that formed the nucleation points that initiate the process. In mammals, anionic surfaces are exposed upon rupture of the **endothelial lining** of the blood vessels and are the binding sites for specific factors that initiate clotting in the intrinsic pathway. Similarly, the term extrinsic came from the observation that there was another factor extrinsic to circulating blood that facilitates blood clotting. This factor was identified as **factor III, tissue factor.** Whether intrinsic or extrinsic, the process of blood coagulation is initiated on the membrane and is continued on the membrane surface at the site of injury.

Hemophilia is an inherited disorder characterized by a permanent tendency for hemorrhages, spontaneous or traumatic, due to a defective blood clotting system. Classic hemophilia, hemophilia A, is an X-linked recessive disorder characterized by a deficiency of factor VIII. About 1 in 10,000 males is born with a deficiency of factor VIII. Of the approximate 25,000 hemophiliacs in the United States, more than 80% are of the A type. Hemophilia B is due to a dysfunction in factor IX.

Some hemophilia A patients may have a normal prothrombin time if the concentration of tissue factor is high. One possible explanation for this is that factor V in human plasma is much lower in concentration than factor X. Activation of an amount of factor X to Xa in excess of that required to bind all of factor Va would initiate blood clotting by the extrinsic pathway and give a normal prothrombin time. The intrinsic pathway would not function normally due to the deficiency in factor VIII. Without the two pathways operating in concert, the overall process of blood clotting would be impaired. Both factor Xa and thrombin activate factor V and are involved in a number of other reactions. If the overall process is not accelerated at its onset by intervention of the intrinsic pathway, due to kinetics of the interaction of thrombin and factor Xa with the normally low concentration of factor V, the clotting disorder is expressed. The blood levels of factor VIII in severe hemophilia A patients are less than 5% of normal. These patients have generally been treated by blood transfusion with its associated dangers: the possibility of contraction of hepatitis or HIV, and the 6% possibility of patients making autoantibodies.

10.5 Blood Coagulation

Calcium ions have at least two important functions in blood coagulation. They form complexes with factors that contain γ-carboxyglutamyl (Gla) residues and induce conformational and electronic states that facilitate their interaction with membrane "receptors" for initiation and localization of their reactions. Calcium ions also bind at sites other than Gla residues, producing protein conformational changes that enhance catalytic activity. Evidence for this second role for calcium ions comes from the observation that activation of at least one of the enzymes leads to both the cleavage and elimination of the N-terminal region containing the Gla residues, but calcium ions are still required for its effective participation in blood coagulation.

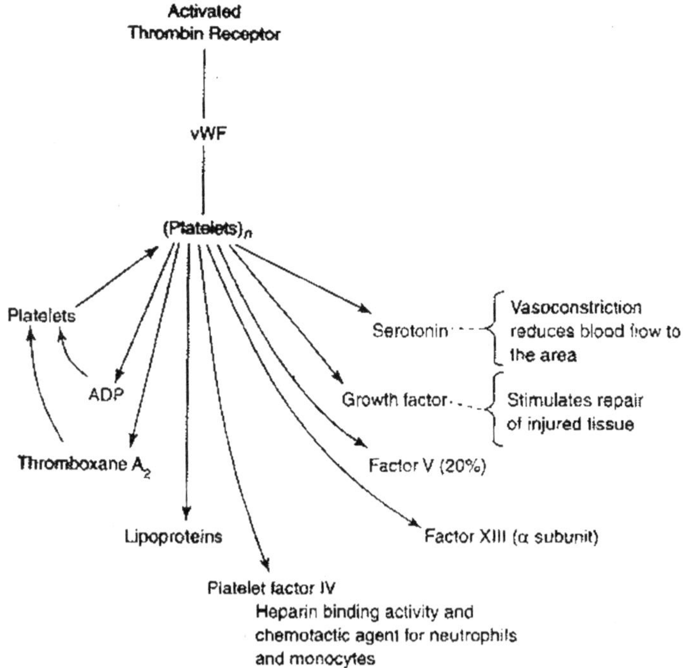

Action of platelets in blood coagulation.

Gla-containing residues are located in the N-terminal region of the molecules followed by a structural component that resembles epidermal growth factor. The position of proteolytic cleavage by activation proteinases is generally at an amino acid residue located

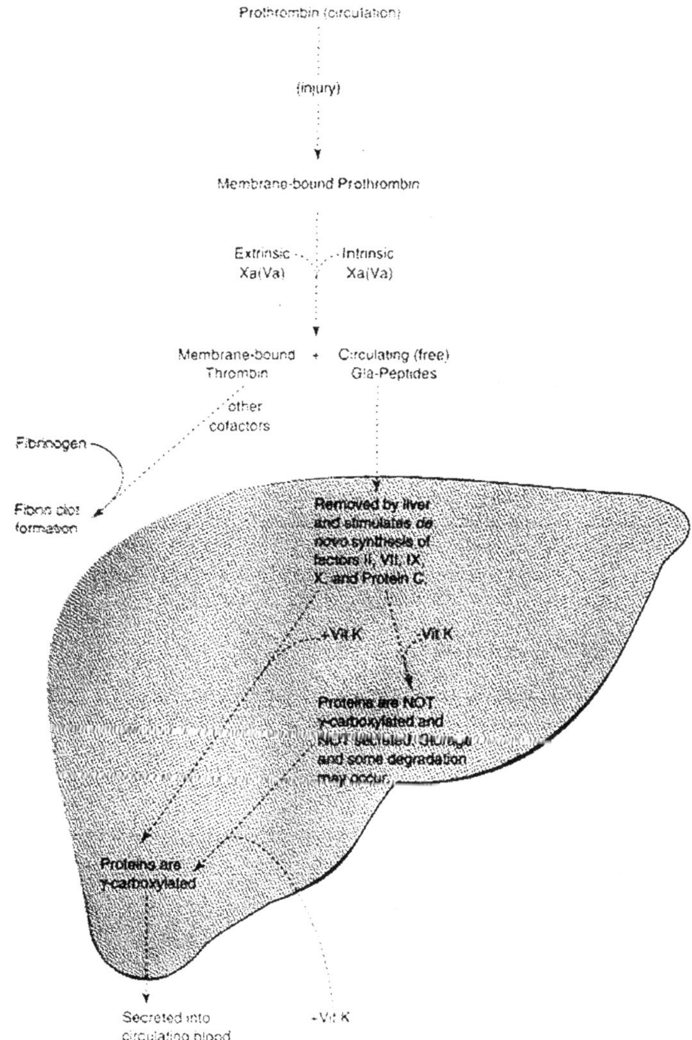

Role of Gla peptides in the regulation of coagulation.

between cysteine residues that form a disulfide bond. Activation may or may not result in loss of a small peptide. *Prothrombin is the only one whose activation is by cleavage outside the bridging disulfide bond and results in elimination of the Gla peptide.* Factor VII is activated by cleavage of a single Arg^{152}-Ile^{153} bond. Factor IX is activated by cleavages at Arg^{145} and Arg^{180} with the release of an approxiamtely 11-kDa peptide. Factor X consists of two chains connected by a disulfide bridge. It is activated by cleavage of its heavy chain at Arg^{194}-Ile^{195}. The Gla residues are located in the light chain. Protein C also consists of a heavy and a light chain connected by a disulfide bond. Cleavage of an Arg-Ile bond at position 169 results in its activation.

Biotechnology of DNA

11.1 The PCR Amplification

The rapid production of large quantities of a specific DNA sequence took a leap forward with the development of the **polymerase chain reaction (PCR).** The PCR requires two nucleotide oligomers that hybridize to the complementary DNA strands in a region of interest. the oligomers serve as primers for a DNA polymerase that extends each strand. Repeated cycling of the PCR yields large amounts of each DNA molecule of interest in a matter of hours as opposed to days and weeks associated with cloning techniques.

The PCR *amplification of a specific DNA sequence* can be accomplished with a purified DNA sample or a small region within a complex mixture of DNA. The nucleotide sequence of the DNA to be amplified must be known or it must be cloned in a vector where the sequence of the flanking DNA has been established. The product of PCR is a double-stranded DNA molecule and the reaction is completed in each cycle when all of the template molecules have been copied. In order to initiate a new round of replication the sample is heated to melt the double-stranded DNA and, in the presence of excess oligonucleotide primers, cooled to permit hybridization of the single-stranded template with free oligomers. A new cycle of DNA replication will initiate in the presence of DNA polymerase and all four dNTPs. Heating to about 95°C as required for melting DNA inactivates most DNA polymerases.

A DNA fragment of unknown sequence is inserted into a vector of known sequence by normal recombination methodologies. The recombinant DNA of interest does not need to be purified from

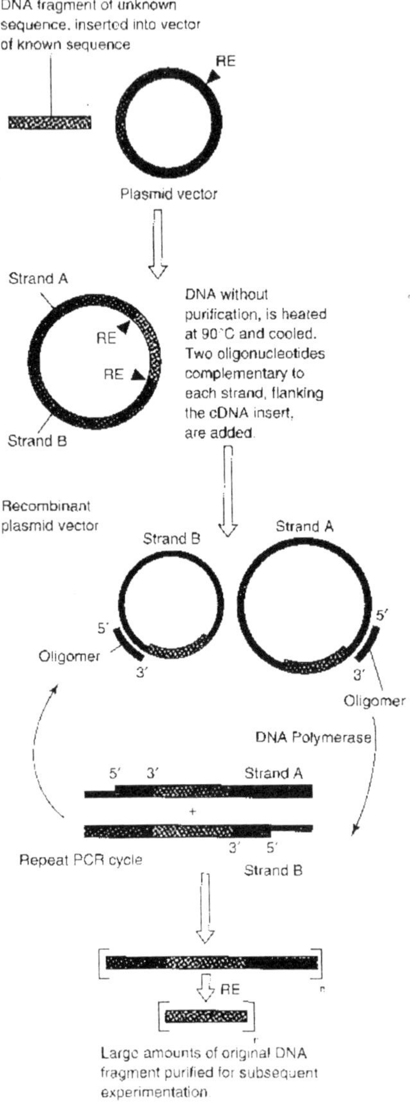

PCR of a DNA.

containing DNA species. The DNA is heated to 90°C to dissociate the double strands and cooled in the presence of excess amounts of two different complementary oligomers that hybridize to the known vector DNA sequences flanking the foreign DNA insert. Only recombinant single-stranded DNA species can serve as templates for DNA replication yielding double-stranded DNA fragments of foreign DNA bounded by the oligomer DNA sequences. The heating-replication cycle is repeated many times to rapidly produce greatly amplified amounts of the original foreign DNA. The DNA fragment of interest can be purified from the polymerase chain reaction mixture by cleaving it with the original restriction endonuclease (RE), electrophoresing the DNA mixture through an agarose gel, and eluting the band of interest from the gel.

Use of the polymerase chain reaction (PCR) to amplify minute quantities of DNA has revolutionized the ability to detect and analyze DNA species. With PCR it is possible to synthesize sufficient DNA for analysis. Conventional methods for detection and identification of the human immunodeficiency virus (HIV), such as Southern blot-DNA hybridization and antigen analysis, are labor intensive, expensive, and have low sensitivity.

11.2 Recombinant DNA Technology

The ability to selectively hydrolyze a population of DNA molecules with a battery of restriction endonucleases led to the development of a technique for joining two different DNA molecules termed **recombinant DNA.** This technique combined with the various techniques for replication, separation, and identification permits the production of large quantities of purified DNA fragments. The combined techniques, referred to as **recombinant DNA technologies,** allow the removal of a piece of DNA out of a large complex molecule, such as the genome of a virus or human, and amplification of the DNA fragment. Recombinant DNAs have been prepared with DNA fragments from bacteria combined with fragments from humans, viruses with viruses, and so on. The ability to join two different pieces of DNA together at specific sites within the molecules is achieved with two enzymes, a restriction endonuclease and a **DNA ligase.** There are a number of different restriction endonucleases, varying in their nucleotide sequence specificity, that can be used. Some hydrolyze the two strands of DNA in a staggered fashion, producing "*sticky* or *cohesive" ends,* while others cut both strands symmetrically, producing a *blunt end.* A specific restriction enzyme cuts DNA at exactly the same nucleotide sequence site regardless of the source of the DNA (bacteria, plant, mammal, etc.). A DNA molecule may have one, several, hundreds, thousands, or no recognition sites for a particular restriction endonuclease. The staggered cut results in a fragmented DNA molecule with ends that are single stranded. When different DNA fragments generated by the same restriction endonuclease are mixed, their single-stranded ends can hybridize, that is, anneal together. In the presence of DNA ligase the two fragments are connected covalently, producing a recombinant DNA molecule.

The DNA fragments produced from restriction endonuclease that form blunt ends can also be ligated but with much lower efficiency. The efficiency can be increased by enzymatically adding a poly(dA) tail to one species of DNA and a poly(dT) tail to the ends of the second species of DNA.

Synthesis of a recombinant DNA open the way for production of significant quantities of interesting DNA fragments. By incorporating a recombinant DNA into a cellular system that allows replication of recombinant DNA, amplication of DNA of interest can be achieved. A carrier DNA, termed a **cloning vector,** is employed. Bacterial plasmids are ideally suited as recombinant DNA vectors. Many bacteria contain a single circular chromosome of approximately 4 million base pairs and minicircular DNA molecules called **plasmids.** Plasmids are usually composed of only a few thousand base pairs and are rarely associated with the large chromosomal molecule. Genes within the plasmid have various functions; one of the most useful is the ability to confer antibiotic resistance to the bacterium, an attribute useful in selecting specific colonies of the bacteira. Plasmids replicate independently of replication of the main bacterial chromosome. One type of plasmid, the **relaxed-control** plasmids, may be present in tens to hundreds of copies per bacterium, and replication is dependent solely on host enzymes that have long half-lives. Therefore replication of **"relaxed" plasmids** can occur in the presence of a protein synthesis inhibitor. Bacteria can accumulate several thousand plasmid copies per cell under these conditions. Other plasmid types are subjected to **stringent control** and their replication is dependent on the continued synthesis of plasmid-encoded proteins. These plasmids replicate at about the same rate as the large bacterial chromosome, and only a low number of copies occur per cell. The former plasmid type is routinely used for recombinant DNA studies.

Once the plasmid has been introduced into the bacterium, both can replicate. Method are available to select those bacteria that carry the recombinant DNA molecules. In the recombinant process some bacteria may not be transformed or may be transformed with a vector not carrying foreign DNA; in preparing the vector some may reanneal without inclusion of the DNA of interest. In some experimental conditions one can generate DNA fragments that can be readily purified for recombinant studies. Such fragments can be generated from small, highly purified DNA species, for example, some DNA viruses. More typically, however, a single restriction endonuclease will generate hundreds to hundreds of thousands of DNA fragments, depending on the size and complexity of DNA being studied. Individual fragments cannot be isolated from these samples to be individually incorporated into the plasmid. Methods have therefore been developed to select those bacteria containing the desired DNA.

Restriction endonucleases do not necessarily hydrolyze DNA into fragments containing intact genes. If the fragment contains an entire gene it may not contain the required flanking regulatory sequences, such as the promoter region. If the foreign gene is of mammalian origin, its regulatory sequences would not be recognized by the bacterial synthetic machinery. The primary gene transcript (pre-mRNA) can also contain introns that cannot be processed by the bacteria.

When a complex mixture of thousands of different genes, arranged on different chromosomes, as in the human genome, is subjected to hydrolysis with a single restriction endonuclease, thousands of DNA fragments are generated. These DNA fragments are annealed with a plasmid vector that has been cleaved to a linear molecule with the same restriction endonuclease. By adjusting the ratio of plasmid to foreign DNA the probability of joining at least one copy of each DNA fragment within a cyclized recombinant-plasmid DNA approaches one. Usually, only one out of the multiple DNA fragment is inserted into each plasmid vector. Bacteria are transformed with the recombinant molecules such that only one plasmid is taken up by a single bacterium. Each recombinant molecule can now be replicated within the bacterium and the bacterium will give rise to progeny, each carrying multiple copies of the recombinant DNA. That total population of bacteria now contain fragments of DNA that may represent the entire human genome. This is termed a **gene library.** As in any library containing thousands of volumes, a selection system must be available to retrieve the book or gene of interest.

Plasmids are commonly employed to clone DNA fragments generated from molecules of limited size and complexity, such as viruses, and to subclone large DNA fragments previously cloned in other vectors. Genomic DNA fragments are usually cloned from other vectors capable of carrying larger foreign DNA fragments than plasmids.

11.3 DNA and RNA Probes

Selection of bacteria harboring recombinant DNA of interest, analysis of mRNAs expressed in a cell, or identification of the presence of DNA sequences within a genome require sensitive and specific detection methods. DNA and RNA **probes** meet these requirements. These probes contain nucleotide sequences complementary to the target nucleic acid and will thus hybridize with the nucleic acid of interest. The degree of complementarity of a probe with the DNA under investigation determines the tightness of binding of the probe. The probe does not need to contain the entire complementary sequence of the DNA. The probe, RNA or DNA, can be labeled, usually with ^{32}P. Nonradioactive labels are also employed that depend on enzyme substrates coupled to nucleotides, which when incorporated into the nucleic acid can be detected by an enzyme-catalyzed reaction.

Labeled probes can be produced by **nick translation** of double-stranded DNA. Nick translation involves the random enzymatic hydrolysis of a phosphodiester bond in the backbone of one strand of DNA by DNase I; the enzymatic breaks in the DNA backbone are referred to as nicks. A second enzyme, *E. coli* DNA polymerase I, with its 5′ → 3′ exonucleolytic activity and its DNA polymerase activity, creates single-strand gaps by hydrolyzing nucleotides from the 5′ side of the nick and then filling in the gaps with its polymerase activity. The polymerase reaction is usually carried out in the presence of one α-^{32}P-labeled deoxynucleotide triphosphate and three unlabeled deoxynucleotide triphosphates. The DNA employed in this method is usually purified and is derived from cloned DNA, viral DNA, or cDNA.

Another method of label DNA probes, **random primer labeling of DNA,** has distinct advantages over the nick translation method. The random primer method typically requires only 25 ng

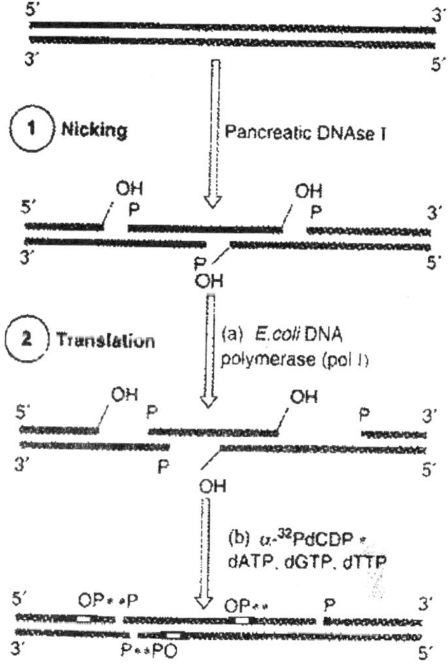

Nick translation to label DNA probes.

of DNA as opposed to 1–2 μg of DNA for nick translation and results in labeled probes with a specific activity (>10^9 cmp μg^{-1}) approximately ten times higher. This method generally produces longer labeled DNA probes. The double-stranded probe is melted and hybridized with a mixture of random hexanucleotides containing all possible sequences (ACTCGG, ACTCGA ACTCGC, etc.). The hybridized hexanucleotides serve as primers for DNA synthesis with a DNA polymerase, such as the Klenow enzyme, in the presence of one or more radioactively labeled deoxynucleoside triphosphates.

Labeled RNA probes have advantages over DNA probes. For one, relatively large amounts of RNA can be transcribed from a template, which may be available in very limited quantities. A double-stranded DNA (dsDNA) probe must be denatured prior to hybridization with the target DNA and rehybridization with itself competes for hybridization with the DNA of interest. No similar competition occurs with the single-stranded RNA probes that hybridize with complementary DNA or RNA molecules. Synthesis of an RNA probe requires DNA as a template. To be transcribed the template DNA must be covalently linked to an upstream promoter that can be recognized by a DNA-dependent RNA polymerase. Vectors have been constructed that are well suited for this technique.

A labeled DNA or RNA probe can be hybridized to nitrocellulose-bound nucleic acids and identified by the detection of the labeled probe.

Purified DNA molecules can be radioactively labeled and used to detect, by hybridization, the presence of complementary RNA or DNA in experimental samples. (1) Nicking step: introduces random single-stranded breaks in the DNA. (2) Translation step: (a) *E. coli* DNA polymerase (pol I) has 5′ → 3′ exonucleolytic activity that hydrolyzes nucleotides from the 5′ end of the nick; (b) pol I simultaneously fills in the single-stranded gap with radioactively labele nucleotides using the 3′ end as a primer.

11.4 Synthsis of cDNA

Messenger RNA can be reverse transcribed on cDNA and the cDNA inserted into a vector for amplification identification, and expression. Mammalian cells normally contain 10,000–30,000 different species of mRNA molecules at any time during the cell cycle. In some cases, however, a specific mRNA species may approach 90% of the total mRNA, such as mRNA for globin in reticulocytes. Many mRNAs are normally present at only a few (1–14) copies per cell. A **cDNA library** can be constructed from the total cellular mRNA but if only a few copies per cell of mRNA of interest are present, the cDNA may be very difficult to identify. Methods that enrich the population of mRNAs or their corresponding cDNAs permit reduction of the number of different cDNA species within a cDNA library and greatly enhance the probability of identifying the clone of interest.

Messenger RNA can be separated by size by gel electrophoresis or centrifugation. Utilization of mRNA in a specific molecular size range will enrich several-fold an mRNA of interest. Knowledge of the molecular weight of the protein encoded by the gene of interest gives a clue to the approximate size of the mRNA transcript or its cDNA; variability in the predicted size, however, will arise from differences in the length of the untranslated regions of the mRNAs.

Enrichment of a specific mRNA molecule can also be accomplished by immunological procedures but requires the availability of antibodies against the protein encoded by the gene of interest. Antibodies added to an *in vitro* protein synthesis mixture will react with the growing polypeptide chain associated with the polysome and precipitate it. The mRNA can be purified from the immunoprecipitated polysomal fraction.

An isolated mRNA mixture is used as a template to synthesize a complementary strand of DNA using RNA-dependent DNA polymerase, reverse transcripted. A primer is required for the reaction; advantage is taken of the poly(A) tail at the 3′ terminus of eukaryotic mRNA. An oligo(dT) with 12–18 bases is employed as the primer that will hybridize with the poly(A) sequence. After cDNA synthesis, the hybrid is denatured or the mRNA hydrolyzed in alkali in order to obtain the single-stranded cDNA. The 3′ termini of single-stranded cDNA form a hairpin loop that serves as a primer for the synthesis of the second strand of the cDNA. Either the Klenow fragment or a reverse transcriptase can be used for this step. The resulting double-stranded cDNA contains a single-stranded loop that is selectively recognized and digested by S1 nuclease. The ends of the cDNA must be modified prior to cloning in a vector. One method involves incubating blunt-ended cDNA molecules with linker molecules and a bacteriophage T4 DNA ligase that catalyzes the ligation of blunt-ended molecules. The synthetic linker molecules contain restriction endonuclease sites that can now be hydrolyzed with the appropriate enzyme for insertion of the cDNA into a compatibly cut vector.

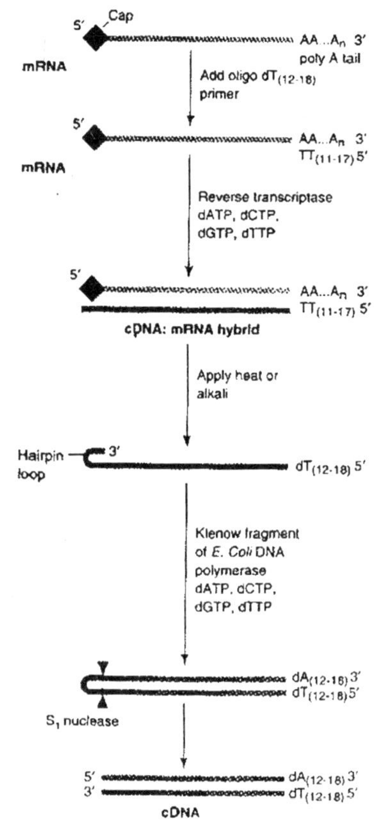

Synthesis of cDNA from mRNA.

Bacteriophage DNA is the most convenient and efficient vector to create cDNA libraries because they can readily be amplified and stored indefinitely two bacteriophage vectors, **λgt10** and **λgt11,** and their newer constructs have been employed to produce cDNA libraries. The cDNA libraries in λgt10 can be screened only with labeled nucleic acid probes, whereas those in λgt11, an expression vector, can also be screened with antibody for the production of the protein or antigen of interest.

11.5 Fusion Protein and Expression Vectors

Recombinant DNA methodology described to this point has dealt primarily with screening, amplification, and purification of cloned DNA species. An important goal of recombinant DNA studies, as stated earlier, is to have a foreign gene expressed in bacteria with the product in a biologically active form. Sequencing the DNA of many bacterial genes and their flanking regions has identified the spatial arrangement of regulatory sequences required for expression of genes. A promoter and other regulatory elements upstream of the gene are required to transcribe a gene. mRNA transcript of a recombinant eukaryotic gene, however, is not translated in a bacterial system because it lacks the bacterial recognition sequence, the Shine-Dalgarno sequence, required to properly orient it with a functional bacterial ribosome. Vectors that facilitate the functional transcription of DNA inserts, termed **expression vectors,** have been constructed such that a foreign gene can be inserted into the vector downstream of a regulated promoter but within a bacterial gene, commonly the *lacZ* gene. The mRNA transcript of the recombinant DNA contains the *lacZ* Shine-Dalgarno sequence, codons for a portion of the 3′ end of the *lacZ* gene protein, followed by the codons of the complete foreign gene of interest. The protein product is a **fusion protein** that contains a few N-terminal amino acids of the *lacZ* gene protein and the complete amino acid sequence of the foreign gene product.

Many plasmid and bacteriophage vectors have been constructed to permit expression of eukaryotic genes in bacterial cells. Rapidly replicating bacteria can serve as a biological factory to produce large amounts of specific proteins, which have research, clinical, and commercial value. As an example, human protein hormones are produced by recombinant technologies, which serve as replacement or supplemental hormones in patients with aberrant or missing hormone production. Depicts a generalized plasmid vector for the expression of a mammalian gene. Recall that the inserted foreign gene must be in the form of cDNA from its corresponding mRNA since the bacterial system cannot remove the introns in the pre-mRNA transcript. The DNA must be inserted in register with the codons of the 3′-terminal codons of the bacterial protein when creating a fusion protein. That is, insertion must occur after a triplet colon of the bacterial protein and at the beginning of a triplet codon of the eukaryotic gene protein to ensure proper translation. Finally, the foreign gene must be inserted in the proper orientation relative to the promoter to yield a functional transcript. This can be achieved by directional cloning.

Eukaryotic proteins synthesized within bacteria are often unstable and are degraded by intracellular proteases. Fusion protein products, however, are usually stable. The fusion protein amino acids encoded by the prokaryotic genome may be cleaved from the purified protein of interest by enzymatic or chemical procedures. An alternative cloning strategy to circumvent the intracellular instability of some proteins is to produce a foreign protein that is secreted. This requires cloning the foreign gene in a vector such that the fusion protein synthesized contains a signal peptide that can be recognized by the bacterial signal peptidase that properly processes the protein for secretion.

Expression of recombinant genes in mammalian cells requires the presence of DNA-controlling elements within the vector that are not necessary in the bacterial system. To be expressed in a eukaryotic cell the cloned gene is inserted in the vector in the proper orientation relative to control elements, including a promoter, polyadenylation signals, and an enhancer sequence. Expression may be improved by the inclusion of an intron. Some of all of these DNA elements may be present in the recombinant gene if whole genomic DNA is used for cloning. A particular cloned fragment generated by restriction endonuclease cleavage, however, may not contain the required controlling elements. A cDNA would not possess these required DNA elements. It is therefore necessary that the expression vector to be used in mammalian cells be constructed such that it contains all of the required controlling elements.

An expression vector can be constructed by insertion of required DNA-controlling elements into the vector by recombinant technologies. Enhancer and promoter elements, engineered into an expression vector, should be recognized by a broad spectrum of cells in culture for the greatest applicability of the vector. Controlling elements derived from viruses with a broad host range are used for this purpose and are usually derived from the **papovavirus, simian virus 40 (SV40), Rous sarcoma virus,** or the **human cytomegalovirus.**

The vector must replicate so as to increase the number of copies within each cell or to maintain copies in daughter cells. The vector therefore is constructed to contain DNA sequences that promote its replication in the eukaryotic cell. This DNA region is usually derived from a virus and is referred to as the origin of replication (Ori). Specific protein factors, encoded by genes engineered into the vector or previously introduced into the host genome, recognize and interact with the ori sequences to initiate DNA replication.

11.6 Role of Flanking Regions in DNA

Site-directed mutagenesis can be carried out in various regions of a DNA sequence including the gene itself or the flanking regions. A simple deletion mutation strategy where the sequence of interest is selectively cleaved with restriction endonuclease, the specific sequence removed, and the altered recombinant vector recircularized with DNA ligase. The role of the deleted sequence can be determined by comparing the level of expression (translation) of the gene product, measured immunologically or enzymatically, to the unaltered recombinant expression vector. A similar technique is used to insert new sequences at the site of cleavage. Deletion of a DNA sequence within the flanking region of a cloned gene can help to define its regulatory role in gene expression. The presence or absence of a regulatory sequence may not be sufficient to evaluate its role in controlling expression. The spatial arrangement of regulatory elements to one another, to the gene, and to its promoter may be important in the regulation of gene expression.

Analysis of potential regulatory sequences is conveniently conducted by inserting the sequence of interest upstream of a reporter gene in an expression vector. A **reporter gene,** usually of prokaryotic origin, encodes for a gene product that can readily be distinguished from proteins normally present in the nontransfected cell and for which there is a convenient and rapid assay. A commonly used reporter gene is the chloramphenicol acetyltransferase (*CAT*) gene of bacteria. The gene product catalyzes the acetylation and inactivation of chloramphenicol, a protein synthesis inhibitor of prokaryotic cells. The ability of a regulatory element to enhance or suppress expression of the *CAT* gene can be determined by assaying the level of acetylation of chloramphenicol in extracts prepared from transfected cells. The regulatory element can be mutated prior to insertion into the vector carrying the reporter gene to determine its spatial and sequence requirements as a regulator of gene expression.

A difficulty encountered in analysis of regulatory elements is the lack of restriction endonuclease sites at useful positions within the cloned DNA. **Deletion mutations** can be made, in the absence of appropriately positioned restriction endonuclease sites, by linearizing cloned DNA with a restriction endonuclease downstream of the potential regulatory sequence of interest. The DNA can then be systematically truncated with an exonuclease, which hydrolyzes nucleotides from the free end of both strands of the linearized DNA. Increasing times of digestion generates smaller DNA fragments. The enzymatic hydrolysis of the double strand of DNA occurs at both ends of the linearized recombinant vector, destroying the original restriction endonuclease site (RE_2). A unique restriction endonuclease site is reestablished to recircularize the truncated DNA molecule for further manipulations to evaluate the function of the deleted sequence. This is accomplished by ligating the blunt ends with a linker DNA, a synthetic oligonucleotide containing one or more restriction endonuclease sites. The ligated linkers are cut with the appropriate enzyme permitting recircularization and ligation of the DNA.

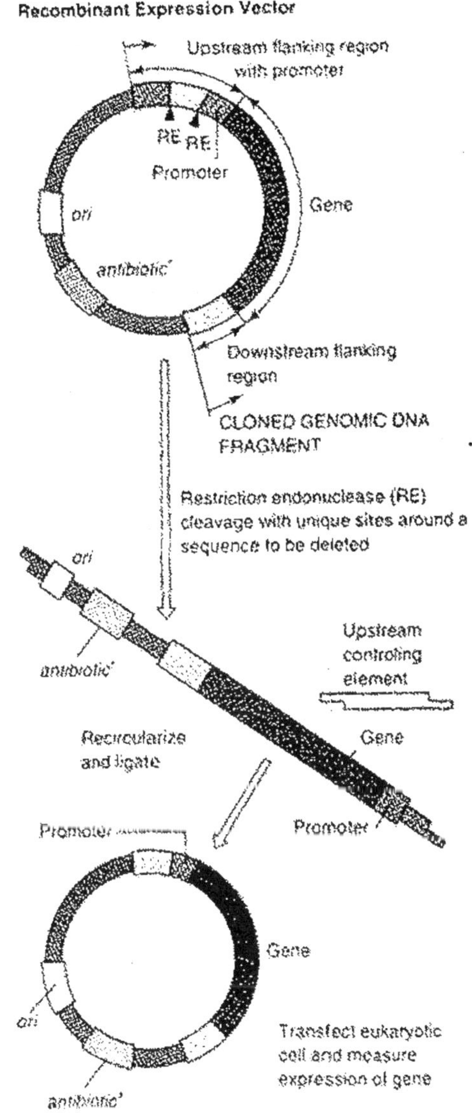

Use of expression vectors to study DNA regulatory sequences.

The gene of interest along with upstream and/or downstream DNA flanking regions is inserted and cloned in an expression vector and the baseline expression of the gene in an appropriate cell is determined. Defined regions of potential regulatory sequences can be removed by restriction endonuclease cleavage and the truncated recombinant DNA vector can be recircularized, ligated, and transfected into an appropriate host cell. The level of gene expression in the absence of the potential regulator is determined and compared to controls to ascertain the regulatory role of the deleted flanking DNA sequence.

12

The Principle of Enzymatic Breakdown of Food

12.1 Introduction

Secretion of digestive fluids and digestion of food were some of the earliest biochemical events to be investigated at the beginning of the era of modern science. Major milestones were the discovery of hydrochloric acid secretion by the stomach and enzymatic hydrolysis of protein and starch by gastric juice and saliva, respectively. The discovery of gastric HCl production goes back to the American physician William Beaumont (1785-1853). In 1822 he treated a patient with a stomach wound. The patient recovered from the wound, but retained a gastric fistula (abnormal opening through the skin). Beaumont seized the opportunity to obtain and study gastric juice at different times during and after meals. Chemical analysis revealed, to the surprise of chemists and biologists, the presence of the inorganic acid HCl. This discovery established the principle of unique secretions into the gastrointestinal tract, which are elaborated by specialized glands.

Soon thereafter, the principle of enzymatic breakdown of food was recognized. In 1836 Theodor Schwann, a German anatomist and physiologist (1810-1882), noticed that gastric juice degraded albumin in the presence of dilute acid. He recognized that a new principle was involved and coined the word pepsin from the Greek *pepsis*, meaning digestion. Today the process of secretion of digestive fluids, digestion of food, and absorption of nutrients and of electrolytes can be described in considerable detail.

The basic nutrients fall into the classes of proteins, carbohydrates, and fats. Many different types of food can satisfy the nutritional needs of humans, even though they differ in the ratios of proteins to carbohydrates and to fats and in the ratio of digestible to nondigestible materials. Unprocessed plant products are especially rich in **fibrous material** that can be neither digested by human enzymes nor easily degraded by intestinal bacteria. The fibers are mostly carbohydrates, such as **cellulose** (β-1, 4-glucan) or **pectins** (mixtures of methyl esters of polygalacturonic acid, polygalactose, and polyarabinose). High-fiber diets enjoy a certain popularity nowadays because of a postulated preventive effect on development of colonic cancer.

The intake of individuals may substantially deviate from the average, as food consumption depends mainly on availability and individual testes. The ability to utilize a wide variety of food is possible because of the great adaptability and digestive reserve capacity of the gastrointestinal tract.

Knowledge of the nature of proteins and carbohydrates in the diet is important from a clinical point of view. Certain proteins and carbohydrates, although good nutrients for most humans, cannot be properly digested by some individuals and produce gastrointestinal ailments. Omission of the offending material and change to another diet can eliminate these gastrointestinal problems. Examples of food constituents that can be the cause of gastrointestinal disorders are **gluten,** one of the protein fractions of wheat, and **lactose,** the disaccharide in milk.

The bulk of ingested nutrients consists of large **polymers** that have to be broken down to **monomers** before they can be absorbed and made available to all cells of the body. The complete process from food intake to absorption of nutrients into the blood consists of a complicated sequence of events.

1. Mechanical homogenization of food and mixing of ingested solids with fluids secreted by the glands of the gastrointestinal tract.
2. Secretion of digestive enzymes that hydrolyze macromolecules to oligomers dimers, or monomers.

3. Secretion of electrolytes, acid, or base to provide an appropriate environment for optimal enzymatic digestion.
4. Secretion of bile acids as detergents to solubilize lipids and facilitate their absorption.
5. Hydrolysis of nutrient oligomers and dimers by enzymes on the intestinal surface.
6. Transport of nutrient molecules and of electrolytes from the intestinal lumen across the epithelial cells into blood for lymph.

To accomplish these functions, the gastrointestinal tract contains specialized glands and surface epithelia:

Organ	*Major Function in Digestion and Absorption*
Salivary glands	Elaboration of fluid and digestive enzymes
Stomach	Elaboration of HCl and proteases
Pancreas	Elaboration of $NaHCO_3$ and enzymes for intraluminal digestion
Liver	Elaboration of bile acids
Gallbladder	Storage and concentration of bile
Small intestine	Terminal digestion of food, absorption of nutrients and electrolytes
Large intestine	Absorption of electrolytes

The **pancreas** and **small intestine** are essential for digestion and absorption of all basic nutrients. Fortunately, both organs have large reserve capacities. For example, maldigestion due to pancreatic failure becomes a problem only when the pancreatic secretion rate of digestive enzymes drops below one-tenth of the normal rate. The secretion of the liver (**bile**) is important for efficient lipid absorption, which depends on the presence of bile acids. In contrast, gastric digestion of food is nonessential for adequate nutrition, and loss of this function can be compensated for by the pancreas and the small intestine. Yet normal gastric digestion greatly increases the smoothness and efficiency of the total digestive process. The stomach aids in the digestion through its reservoir function, its churning ability, and initiation of protein hydrolysis, which, although small, is important for stimulation of pancreatic and gallbladder output. Peptides and amino acids liberated in the stomach stimulate the coordinated release of pancreatic juice and bile into the lumen of the small intestine, thereby ensuring efficient digestion of food.

12.2 The Mode of Solute Transport

Solute movement across an epithelial cell layer is determined by the properties of epithelia cells, particularly their plasma membranes, and by the intercellular tight junctional complexes. The **tight junctions** extend in a belt-like manner around the perimeter of each epithelial cell and connect neighboring cells. Therefore the tight junctions constitute part of the barrier between the two extracellular spaces on either side of the epithelium, that is, the lumen of the gastrointestinal tract and the intercelluler (interstitial) space on the other (blood or serosal) side. The tight junction marks the boundary between the luminal and contraluminal region of the plasma membrane of epithelial cells.

Two potentially parallel pathways for **solute transport** across epithelial cell layers can be distinguished: through the cells (**transcellular**) and through the tight junctions between cells **(paracellular).** The transcellular route in turn consists mainly of two barriers in series, formed by the luminal and contraluminal plasma membranes. Because of this combination of different barriers in parallel (cellular and paracellular pathways) and in series (luminal and contraluminal plasma membranes), biochemical and biophysical information on all three barriers as well as their mutual influence is required for understanding the overall transport properties of the epithelium.

A major function of gastrointestinal epithelial cells is **active transport** of nutrients, electrolytes, and vitamins. The cellular basis for this vectorial solute movement lies in the different properties of the luminal and contraluminal regions of the plasma membrane. The small intestinal cells provide a prominent example of the differentiation and specialization of the two types of membrane. The luminal and contraluminal plasma membranes differ in morphological appearance, enzymatic composition, chemical composition, and transport functions. The luminal membrane is in contact with the nutrients in the chyme (the semifluid mass of partially digested food) and is specialized for terminal digestion of nutrients through its digestive enzymes and for nutrient absorption through transport systems that accomplish concentrative uptake. Transport systems are present for monosaccharides, amino acids, peptides, and electrolytes. In contrast, the contraluminal plasma membrane, which is in contact with the intercellular fluid, capillaries, and lymph, has properties similar to the plasma membranes of most cells.

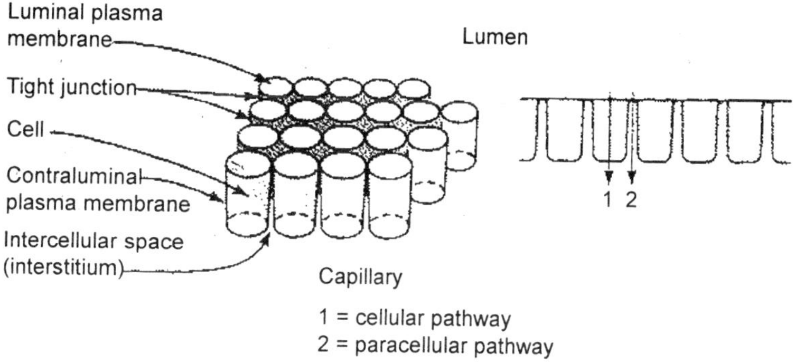

Pathways for transport across epithelia.

12.3 Protein Digestion

The total daily protein load to be digested consists of about 70-100 g of dietary proteins and 35-200 g of endogenous proteins from digestive enzymes and sloughed-off cells. Digestion and absorption of proteins are very efficient processes in healthy humans, since only about 1-2 g of nitrogen are lost through feces each day, which is equivalent to 6-12 g of protein.

Except for a short period after birth, oligo- and polypeptides (proteins) are not absorbed intact in appreciable quantities by the intestine. Proteins are broken down by hydrolases with specificity for the peptide bond, that is, by peptidases. This class of enzymes

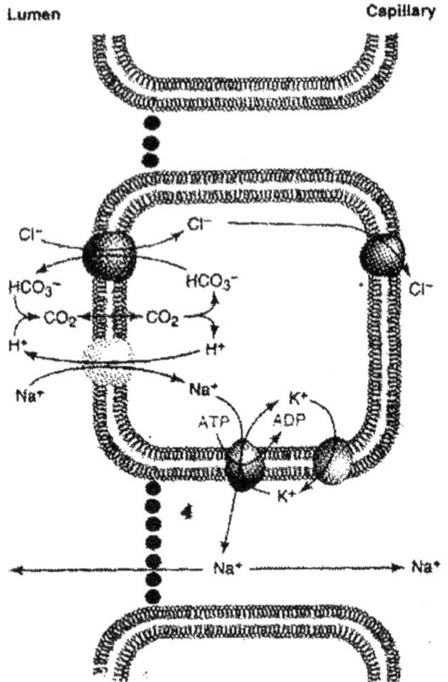

Model for electrically neutral NaCl absorption in the small intestine.

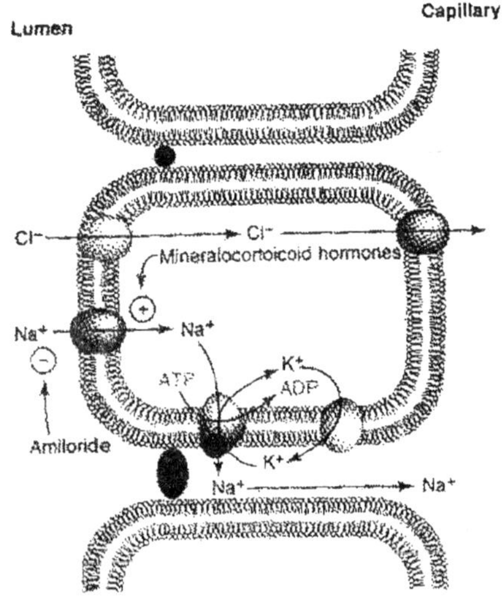

Model for electrogenic NaCl absorption in the lower intestine.

is divided into **endopeptidases** (proteases), which attack internal bonds and liberate large peptide fragments, and **exopeptidases,** which cleave off one amino acid at a time from either the COOH **(carboxypeptidases)** of the NH_2 terminus **(aminopeptidases).** Endopeptidases are important for an initial breakdown of long polypeptides into smaller products, which can then be attacked more efficiently by exopeptidases. The final products are free amino acids and di- and tripeptides, which are absorbed by epithelial cells.

The process of protein digestion can be divided into a gastric, a pancreatic, and an intestinal phase, depending on the source of peptidases.

Gastric juice is characterized by the presence of HCl and therefore a low pH less than 2 as well as the presence of proteases of the pepsin family. The acid serves to kill off microorganisms and also to **denature proteins.** Denaturation makes proteins more susceptible to hydrolysis by proteases. **Pepsins** are unique in that they are acid stable; in fact, they are active at acid but not at neutral pH. The catalytic mechanism that is effective for peptide hydrolysis at the acid pH depends on two carboxylic groups at the active site of the enzymes. Pepsin A, the major gastric protease, prefers peptide bonds formed by the amino group of aromatic acids (Phe, Tyr).

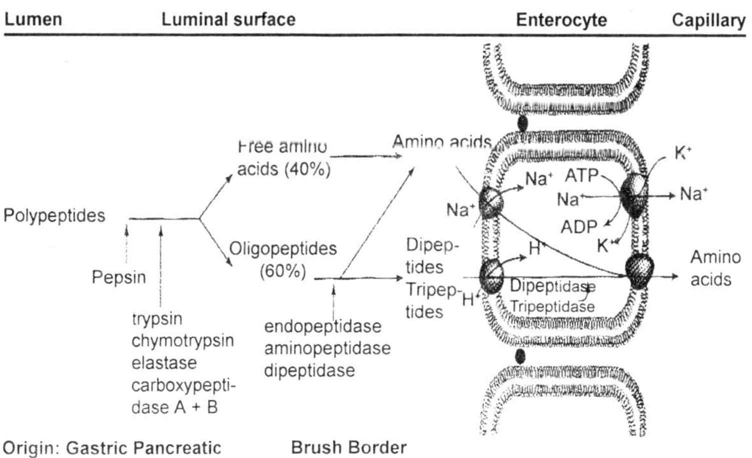

Digestion and absorption of proteins.

Active pepsin in generated from the proenzyme **pepsinogen** by the removal of 44 amino acids from the NH_2 terminus (pig enzyme). Cleavage between residues 44 and 45 of pepsinogen occurs as either an intramolecular reaction **(autoactivation)** below pH 5 or by active pepsin (autocatalysis). The liberated peptide from the NH_2 terminus remains bound to pepsin and acts as "pepsin inhibitor" above pH 2. This inhibition is released either by a drop of the pH below 2 or further degradation of the peptide by pepsin. Thus, once favorable conditions are reached, pepsinogen is converted to pepsin by autoactivation and subsequent autocatalysis at an exponential rate.

The major products of pepsin action are large peptide fragments and some free amino acids. The importance of gastric protein digestion does not lie so much in its contribution to the breakdown of ingested macromolecules, but rather in the generation of peptides and amino acids that act as stimulants for **cholecystokinin** release in the duodenum. The gastric peptides therefore are instrumental in the initiation of the pancreatic phase of protein digestion.

Pancreatic juice is rich in proenzymes of **endopeptidases** and **carboxypeptidases** which are activated after they reach the lumen of the small intestine. **Enteropeptidase** (old name: enterokinase), as protease produced by duodenal epithelial cells, activates pancreatic **trypsinogen** to **trypsin** by scission of a hexapeptide from the NH_2 terminus. Trypsin in turn autocatalytically activates more trypsinogen to trypsin and also acts on the other proenzymes, thus liberating the endopeptidases chymotrypsin and elastase and the **carboxypeptidases A and B.** Since trypsin plays a pivotal role among pancreatic enzymes in the activation process, pancreatic juice normally contains a small-molecular-weight peptide that acts as a **trypsin inhibitor** and neutralizes any trypsin formed prematurely within the pancreatic cells or pancreatic ducts.

Trypsin, chymotrypsin, and elastase have different substrate specificity. They are active only at neutral pH and depend on pancretic $NaHCO_3$ for neutralization of gastric HCl. Their mechanism of catalysis involves an **essential serine residue** and is thus similar to serine esterases, such as acetyl choline esterase.

Reagents that interact with serine and modify it, inactivate serine esterases and peptidases. A prominent example of such a reagent is the highly toxic diisoprophylphosphofluoridate, which was developed originally for chemical warfare (neurotoxic because of inhibition of acetyl choline esterase).

12.4 Digestion and Absorption of Carbohydrates

Dietary carbohydrates provide a major portion of the daily caloric requirement. They consist of mono-, di-, and poly-saccharides. **Monosaccharides** need not be hydrolyzed for absorption. Disaccharides require the small intestinal surface enzymes for hydrolysis into monosaccharides, while polysaccharides depend on **pancreatic amylase** for degradation.

Starch, a major nutrient, is a plant polysaccharide with a molecular mass of more than 100 kDa. It consists of a mixture of linear chains of glucose molecules linked by α-1,4-glucosidic bonds **(amylose)** and of branched chains with branch points made up by α-1,6, linkages **(amylopectin).** The ratio of 1,4- to 1,6-glucosidic bonds is about 20:1 **Glycogen** is an animal polysaccharide similar in structure to amylopectin. The two compounds differ in terms of the number of branch points, which occur more frequently in glycogen.

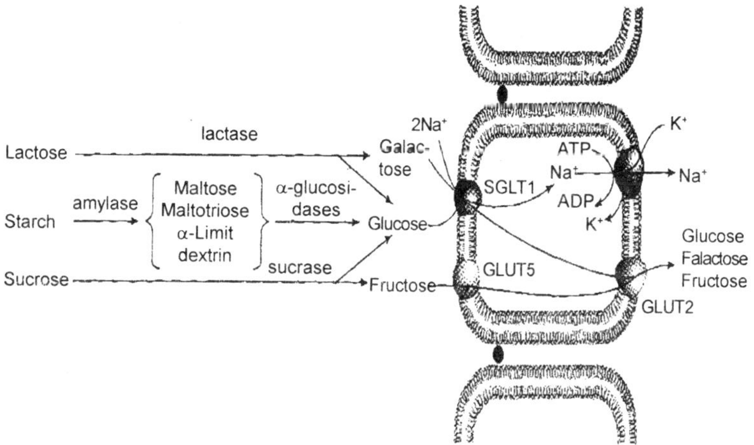

Digestion and absorption of carbohydrates.

The products of the digestion by α-amylase are mainly the **disaccharide maltose,** the **trisaccharide maltotriose,** and so-called **α-limit dextrins** containing on average eight glucose units with one or more α-1,6-glucosidic bonds.

Final hydrolysis of di- and oligosaccharides to monosaccharides is carried out by surface enzymes of the small intestinal epithelial cells. Most of the surface oligosaccharidases are exoenzymes that cleave off one monosaccharide at a time from the nonreducing end. The capacity of the **α-glucosidases** is normally much greater than that needed for completion of the digestion of starch. Similarly, there is usually excess capacity for sucrose (table sugar) hydrolysis relative to dietary intake. In contrast, **β-galactosidase (lactase)** can be rate-limiting in humans for hydrolysis and utilization of lactose, the major milk carbohydrate.

Di-, oligo-, and polysaccharides that are not hydrolyzed by α-amylase and/or intestinal surface enzymes cannot be absorbed; therefore they reach the lower tract of the intestine, which from the lower ileum on contains bacteria. Bacteria can utilize many of the remaining carbohydrates because they possess many more types of saccharidases than humans. Monosaccharides that are released as a result of **bacterial enzymes** are predominantly metabolized anaerobically by the bacteria themselves, resulting in degradation products such as short-chain fatty acids, lactate, hydrogen gas (H_2), methane (CH_4), and carbon.

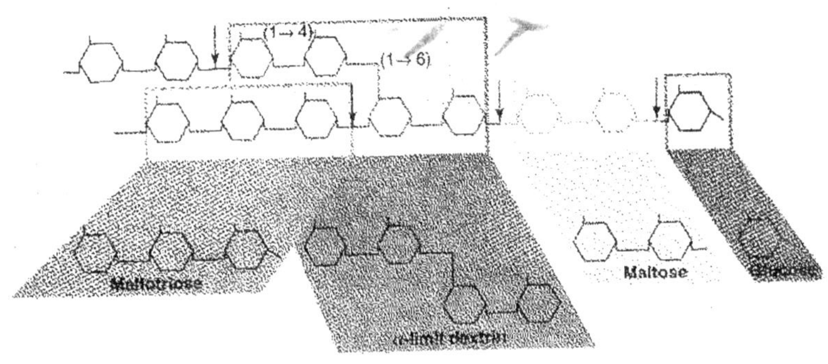

Digestion of amylopectin by salivery and pancreatic *α*-amylase.

Intestinal disaccharidase deficiencies are encountered relatively frequently in humans. Deficiency can be present in one enzyme or several enzymes for a variety of reasons (genetic defect, physiological decline with age, or the result of "injuries" to the mucosa). Of the disaccharidases, lactase is the most common enzyme with an absolute or relative deficiency, which is experienced as milk intolerance. The consequences of an inability to hydrolyze lactose in the upper small intestine are inability to absorb lactose and bacterial fermentation of ingested lactose in the lower small intestine. Bacterial fermentation results in the production of gas (distension of gut and flatulence) and osmotically active solutes that draw water into the intestinal lumen (diarrhea). The lactose in yogurt has already been partially hydrolyzed during the fermentation process of making yogurt. Thus individuals with lactase deficiency can often tolerate yogurt better than unfermented dairy products. The enzyme lactase is commercially available to pretreat milk so that the lactose is hydrolyzed.

12.5 Bile Acid Metabolism

All bile acids are synthesized within the liver from cholesterol but can be modified by bacterial enzymes in the intestinal lumen. **Primary bile acids** synthesized by the liver are **cholic** and **chenodeoxycholic** (or chenic) acid. The **secondary bile acids** are derived from the primary bile acids by bacterial dehydroxylation in position 7 of the ring structure, resulting in **deoxycholate** and **lithocholate,** respectively.

Primary and secondary bile acids are reabsorbed by the intestine into the portal blood, taken up by the liver, and then resecreted into bile. Within the liver, primary as well as secondary bile acids are linked to either glycine or taurine via an isopeptide bond.

These derivatives are called **glyco-** and **tauroconjugates,** respectively, and constitute the forms that are secreted into bile. With the conjugation, the carboxyl group of the unconjugated acid is replaced by an even more polar group. The pK values of the carboxyl group of glycine and of the sulfonyl group of taurine are lower than that of unconjugated bile acids, so that conjugated bile

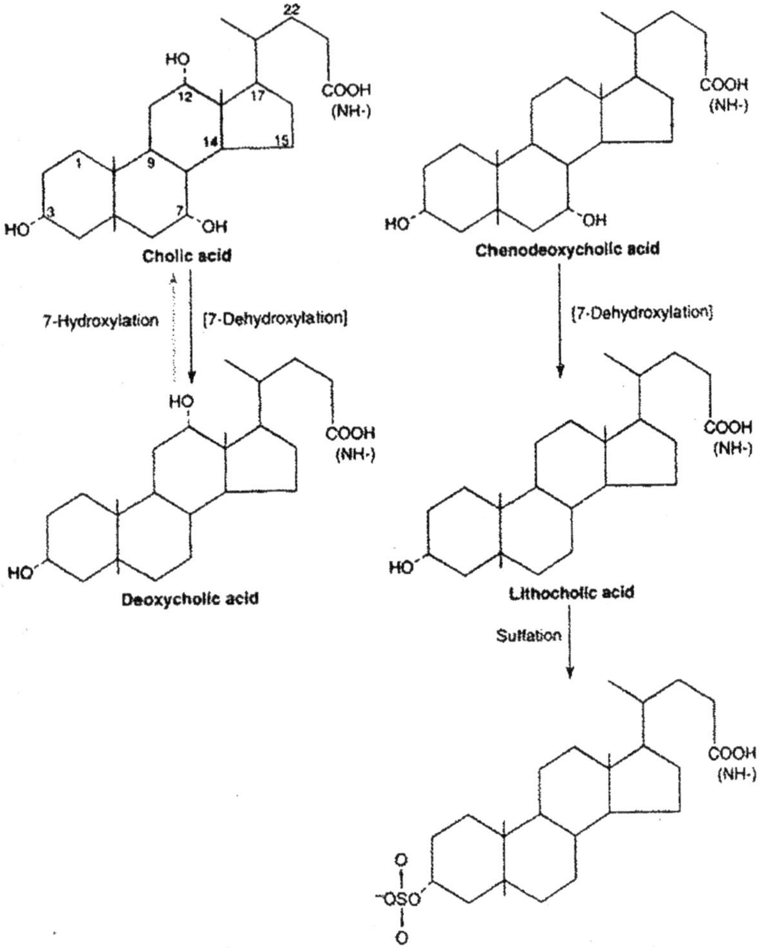
22
HO
12
17
COOH
(NH-)
14
15
1
9
3
7
HO
OH
Cholic acid
7-Hydroxylation
[7-Dehydroxylation]
HO
COOH
(NH-)
HO
Deoxycholic acid
COOH
(NH-)
HO
OH
Chenodeoxycholic acid
[7-Dehydroxylation]
COOH
(NH-)
HO
Lithocholic acid
Sulfation
COOH
(NH-)
O
-OSO
O

Bile aid Metabolism in at.

acids remains ionized over a wider pH range. The conjugation is partially reversed within the intestinal lumen by hydrolysis of the isopeptide bond.

The total amount of conjugated and unconjugated bile acids secreted per day by the liver is 16-70 g for an adult.

13

Macronutrients

13.1 Energy Metabolism

Nutrition is best defined as the utilization of foods by living organisms. Since the process of food utilization is biochemical, the major thrust of the next two chapters is a discussion of basic nutritional concepts in biochemical terms. Simply understanding basic nutritional concepts is no longer sufficient. Nutrition attracts more than its share of controversy in our society, and a thorough understanding of nutrition almost demands an understanding of the issues behind these controversies.

Study of human nutrition can be divided into three areas: undernutrition, overnutrition, and ideal nutrition. **Undernutrition** is not a primary concern in this country because nutritional deficiency diseases are now quite rare. **Overnutrition** is a particularly serious problem in developed countries. Current estimates suggest that between 15% and 30% of the U.S. population is obese, and obesity is known to have a number of serious health consequences. Finally there is increasing interest today in the concept of ideal or **optimal nutrition.** This is a concept that has meaning only in an affluent society. Only when food supply becomes abundant enough so that deficiency diseases are a rarity does it become possible to consider long-range effects of nutrients on health. This is probably the most exciting area of nutrition today.

You should be well acquainted with the energy requirements of the body. Much of the food we eat is converted to ATP and other high-energy compounds, which are utilized to drive biosynthetic pathways, generate nerve impulses, and power muscle contraction. We generally describe the energy content of foods in terms of **calories.** Technically speaking, we are actually referring to **kilocalories** of heat energy released by composition of that food in the body. Some nutritionists prefer the term kilojoule (a measure of mechanical energy), but since the American public is likely to be counting calories rather than joules in the foreseeable future, we will restrict ourselves to that term. Caloric values of protein, fat, carbohydrate, and alcohol are roughly 4, 9, 4 and 7 kcal g^{-1}, respectively. Given these data and the composition of the food, it is simple to calculate the caloric content (input) of the foods we eat. Calculating caloric content of foods does not appear to be a major problem in this country. Millions of Americans are able to do it with ease. The problem lies in balancing caloric input with caloric output. Where do these calories go?

There are four principal factors that affect individual energy expenditure: surface area (which is related to height and weight) age, sex, and activity level. (1) The effects of surface area are thought to be simply related to the rate of heat loss by the body—the greater the surface area, the greater the rate of heat loss. While it may seem surprising, a lean individual actually has a greater surface area, and thus a greater energy requirement than an obses individual of the same weight. (2) Age may reflect two factors: growth and lean muscle mass. In infants and children more energy expenditure is required for rapid growth, and this is reflected in a higher **basal metabolic rate** (rate of energy utilization in resting state). In adults (even lean adults), muscle tissue is gradually replaced with fat and water during the aging process, resulting in a 2% decrease in basal metabolic rate (BMR) per decade of adult life. (3) As for sex, women tend to have a lower BMR than men due to a smaller percentage of lean muscle mass and the effects of female hormones on metabolism. (4) The effect of activity levels on energy requirements is obvious. However, most of us overemphasize the immediate, as opposed to the long-term, effects of exercise. For example, one would need to jog for over an hour to burn up the calories found in one piece of apple pie.

Yet, the effect of a regular **exercise** program on energy expenditure can be quite beneficial. Regular exercise increases lean muscle mass, which has a higher basal metabolic rate than adipose tissue allowing one to burn up calories more rapidly 24 hours a day. A regular exercise program should be designed to increase lean muscle mass and should be repeated 3-5 days a week but need not be aerobic exercise to have an effect on basal metabolic rate. For an elderly or infirm individual, even daily walking may, with time, help to increase basal metabolic rate slightly.

Hormone levels are important also, since thyroxine, sex, hormones, growth hormone, and, to a lesser extent, epinephrine and cortisol increase BMR. The effects of epinephrine and cortisol probably explain in part why severe stress and major trauma significantly increase energy requirements. Finally, energy intake itself has an inverse relationship to expenditure in that during periods of **starvation** or semistarvation BMR can decrease up to 50%. This is of great survival value in cases of genuine starvation, but not much help to the person who wishes to lose weight on a calorie-restricted diet.

13.2 Protein Requirement

One of the most important problems of a purely vegetarian diet (as opposed to a lacto-ovo vegetarian diet) is the difficulty in obtaining sufficient calories and protein. Potential caloric deficit results from the fact that the caloric densities of fruits and vegetables are much less than the meats they replace (30-50 cal per 100 g versus 150-300 cal per 100 g). The protein problem is generally threefold: (1) most plant products contain much less protein (1-2 g of protein per 100 g versus 15-20 g per 100 g); (2) most plant protein is of low biological value; and (3) some plant proteins are incompletely digested. Actually, well-designed vegetarian diets usually provide enough calories and protein for the average adult. In fact, the reduced caloric intake may well be of benefit because strict vegetarian do tend to be lighter than their nonvegetarian counterparts.

However, whereas an adult male may require about 0.8 g of protein and 40 cal kg^{-1} of body weight, a young child may require

2-3 times that amount. Similarly, a pregnant woman needs an additional 10 g of protein and 300 cal day $^{-1}$ and a lactating woman an extra 15 g of protein and 500 cal. Thus both young children and pregnant and lactating women run a risk of protein–energy malnutrition. Children of vegetarian mothers generally have a lower birth weight than children of mothers consuming a mixed diet. Similarly, vegetarian children generally have a slower rate of growth through the first 5 years, but generally catch up by age 10.

It is possible to provide sufficient calories and protein even for these high-risk groups provided the diet is adequately planned. Three principles should be followed to design a calorie-protein-sufficient vegetarian diet for young children: (1) whenever possible, include eggs and milk in the diet; they are both excellent sources of calories and high-quality protein; (2) include liberal amounts of those vegetable food with high-caloric density in the diet, including nuts, grains, dried beans, and dried fruits; and (3) include liberal amounts of high-protein vegetable foods that have complementary amino acid patterns. It used to be thought that these complementary proteins must be present in the same meal. Recent animal studies, however, suggests that a meal low in (but not devoid of) an essential amino acid may be supplemented by adding the limiting amino acid at a subsequent meal.

Chronic renal failure is characterized by the buildup of the end products of protein catabolism, mainly urea. Some degree of dietary protein restriction is usually necessary because these toxic end products are responsible for many of the symptoms associated with renal failure. The amount of protein restriction is dependent on the severity of the disease. It is easy to maintain patients in nitrogen balance for prolonged periods on diets containing as little as 40 g of protein/day if the diet is calorically sufficient. Diets containing less than 40 g/day pose problems. Protein turnover continues and a balance must be found between enough protein to avoid negative nitrogen balance and little enough to avoid buildup of waste products.

Moreover, such diets are extremely monotonous and difficult to follow. A typical 20-g protein diet is shown below:

1. One egg plus ¾ cup milk or 1 additional egg or 1 oz of meat.
2. One-half pound of deglutenized (low-protein) wheat bread; all other breads and cereals must be avoided—this includes almost all baked goods.
3. A limited amount of low-protein, low-potassium fruits and vegetables.
4. Sugars and fats to make up the rest of the needed calories; however, cakes, pies, and cookies need to be avoided.

The palatability of these diets can be improved considerably by starting with a vegan diet and supplementing it with a mixture of essential amino acids and ketoacid analogs of the essential amino acids. Recent studies indicate that this technique will help preserve renal function and allow a somewhat greater variety of foods.

The strategy employed in such diets is twofold: (1) provide a minimum of protein, primarily protein of high BV, and (2) provide the rest of the daily calories as carbohydrate and fats. The goal is to provide just enough essential amino acid to maintain positive nitrogen balance. In turn, the body should be able to synthesize the nonessential amino acids from other nitrogen-containing metabolites. Enough carbohydrate and fat are provided so that essentially all dietary protein can be spared from energy metabolism. With this type of diet, it is possible to maintain a patient on 20 g of protein per day for considerable periods. Because of the difficulty in maintaining nitrogen equilibrium at such low-protein intakes, the patient's protein status should be monitored. This can be done by measuring parameters such as serum albumin and transferring.

13.3 Energy Intake

Much has been said in recent years about the large amount of protein that the average American consumes. Certainly most consume far more than needed to maintain positive nitrogen balance. An average American currently consumes 99 g of protein, 68% from animal sources. However, most studies show that a healthy adult can consume that amount of protein with no apparent harm. Concern has been raised about possible effects of high-protein intake on calcium requirements. Some studies suggest that high-protein intake increases urinary loss of calcium and may accelerate bone demineralization associated with aging. However, this issue is far from settled.

Perhaps the more serious nutritional problem is excessive energy consumption. In fact, **obesity** is the most frequent nutritional disorder in the United States. It would, however, be unfair to label obesity as simply a problem of excess consumption. Overeating plays an important role in many individuals, as does inadequate exercise, but there is also a strong genetic component as well. While the biochemical mechanisms for this genetic predisposition are unclear, investigators have recently identified an obesity gene in mice that appears to regulate obesity through effects of both appetite and deposition of fat. A similar gene exists in humans, but its metabolic function is still not known. Detailed characterization of this and other genes that predispose to obesity in animals may yield valuable clues to the causes and treatment of obesity in humans.

A discussion of the treatment of obesity is clearly beyond the scope of this chapter, but it is worthwhile to consider some of the metabolic consequences of obesity. One striking clinical feature of overweight individuals is a marked elevation of serum free fatty acids, cholesterol, and triacylglycerols irrespective of the dietary intake of fat. Why is this? Obesity it obviously associated with an increased number and/or size of adipose cells. These cells contain fewer **insulin receptors** and thus respond more poorly to insulin, resulting in increased activity of the **hormone-sensitive lipase.** The increased lipase activity along with the increased mass of adipose tissue is probably sufficient to explain the increase in free fatty acids.

These excess fatty acids are carried to the liver and metabolized to acetyl CoA, a precursor for triacylglyceol and cholesterol synthesis. Excess tricylglycerol and cholesterol are released as **very low density lipoprotein particles,** leading to higher circulating levels of both triacylglycerol and cholesterol.

A second striking finding in obses individuals is higher fasting blood sugar levels and decreased glucose tolerance. Fully 80% of **adult-onset diabetics** are overweight. Again the culprit appears to be the decrease in insulin receptors, since many adult-onset diabetics have higher than normal insulin levels. This hyperinsulinemia appears to stimulate the sympathetic nervous system, leading to sodium and water retention and vasoconstriction, which tend to increase blood pressure. Because of these metabolic changes, obesity is a primary risk factor in coronary heart disease, hypertension, and diabetes. This is nutritionally significant because all of these metabolic changes are reversible. Quite often reduction to ideal weight is the single important aim of nutritional therapy. Furthermore, when the individual is at ideal body weight, the composition of the diet becomes a less important consideration in maintaining normal serum lipid and glucose levels.

Any discussion of weight reduction regimens should include a mention of one other metabolic consequence of obesity. As discussed above, obesity can lead to increased retention of both sodium and water. As the fat stores are metabolized, they produce water (which is dense than the fat), and the water may largely be retained. In fact, some individuals may actually observe short-term weight gain on certain diets, even though the diet is working perfectly well in terms of breaking down their adipose tissue. This metabolic fact of life can be psychologically devastating to dieters, who expect quick results for all their sacrifice.

13.4 Metabolic Role of Carbohydrates

The chief metabolic role of carbohydrates in the diet is for energy production. Any carbohydrate in excess of that needed for energy is converted to glycogen and triacylglycerol for long-term storage. The body can adapt to a wide range of carbohydrate levels in the diet. Diets high is carbohydrate result in higher steady-state levels of glucokinase and some of the enzymes involved is the hexose monophosphate shunt and triacylglycerol synthesis. Diets low in carbohydrate result in higher steady-state levels of some of the enzymes involved in gluconcogenesis, fatty acid oxidation, and amino acid catabolism. **Glycogen stores** are also affected by the carbohydrate content of the diet.

The most common nutritional problems involving carbohydrates are seen in those individuals with various **carbohydrate intolerances.** The most common form of carbohydrate intolerance is **diabetes mellitus,** caused either by lack of insulin production or lack of insulin receptors. This causes an intolerance to glucose and sugars that can readily be converted to glucose. **Lactase insufficiency** is also a common disorder of carbohydrate metabolism affecting over 30 million people in the United States alone. It is most prevalent among blacks. Asians, and Hispanics. Without the enzyme lactase, the lactose is not significantly hydrolyzed or absorbed. It remains in the intestine where it acts osmotically to draw water into the gut and serves as a substrate for conversion to lactic acid, CO_2, and H_2S by intestinal bacteria. The end result is bloating, flatulence, and diarrhea—all of which can be avoided simply by eliminating milk and milk products from the diet.

Triacylglycerols, or fats, are directly utilized by many tissues of the body as an energy source and, as phospholipids, are an important part of membrane structure. Excess fat in the diet can only be stored as triacylglycerol. As with carbohydrate, the body adapts to a wide range of fat intakes. However, problems develop at the extremes (either high or low) of fat consumption. At the low end, **essential fatty acid (EFA)** deficiencies may become a problem. The fatty acids linoleic, linolenic, and arachidonic acid cannot be made by the body and thus are essential components of the diet. These EFAs are needed for maintaining the function and integrity of membrane structure, for fat metabolism and transport, and for synthesis of **prostaglandins.** The most characteristic symptom of essential fatty acid deficiency is a scaly dermatitis. EFA deficiency is very rare in the United States, occurring primarily in low-birth-weight infants fed artificial formulas lacking EFA and in hospitalized patients maintained on total parenteral nutrition for long periods of time. At the high end of the scale, there is concern that excess dietary fat causes elevation of serum lipids and thus an increased risk of heart disease. Recent studies also suggest that high-fat intakes are associated with increased risk of colon, breast, and prostate cancer, but it is not yet certain whether the cancer risk is associated with fat intake *per se* or with the excess calories associated with a high-fat diet. To the extent that fat intake is associated with cancer risk, animal studies suggest that polyunsaturated fatty acids of the ω-6 series may be more tumorigenic than other unsaturated fatty acids. The reason for this is not known, but it has been suggested that prostaglandins derived from the ω-6 fatty acids may stimulate tumor progression.

13.5 Cholesterol and Heart Disease

With respect to heart disease, the current discussion centers around two key issues: (1) Can serum cholesterol and triacylglycerol levels be controlled by diet? (2) Does lowering serum cholesterol and triacylglycerol levels protect against heart disease? The controversies centered around dietary control of cholesterol levels illustrate perfectly the trap one falls into by trying to look too closely at each individual component of the diet instead of the diet as a whole. For example, there are at least four dietary components that can be identified as having an effect on serum cholesterol: cholesterol itself, **polyunsaturated fatty acids (PUFAs), saturated fatty acids (SFAs),** and fiber. It would seem that the more cholesterol one eats, the higher the serum cholesterol should be. However, cholesterol synthesis is tightly regulated via a feedback control at the hydroxymethylglutaryl-CoA reductase step, so decreases in dietary cholesterol have relatively little effect on serum cholesterol levels.

While the effects of various lipids in the diet can be dramatic, the biochemistry of their action is still uncertain. Saturated fats inhibit receptor-mediated uptake of LDL, but the mechanism is complex. Palmitic acid (saturated, C16) raises cholesterol levels while stearic acid (saturated, C18) is neutral. Polyunsaturated fatty acids lower both LDL and HDL cholesterol levels, while oleic acid (monounsaturated, C18) appears to lower LDL without affecting HDL levels. Furthermore, the ω-3 and ω-6 polyunsaturated fatty acids have slightly different effects on lipid profiles. However, these mechanistic complexities do not significantly affect dietary recommendations. Most foods high in saturated fats contain both palmitic and stearic acid and are atherogenic. The data showing oleic acid lowers LDL levels means that olive oil, and possibly peanut oil, may bė considered as beneficial as polyunsaturated oils.

There is very little disagreement with respect to these data. The question is, what can be done with information? Much of the disagreement arises from the tendency to look at each dietary factor in isolation. For example, it is debatable whether it is worthwhile placing a patient on a highly restrictive 300-mg cholesterol diet (1 egg = 213 mg of cholesterol) if his serum cholesterol is lowered by only 5–10%. Likewise, changing the **PUFA/SFA ratio** from 0.3 (the current value) to 1.0 would either require a radical change in the diet by elimination of foods containing saturated fat (largely meats and fats) or an addition of large amounts of rather unpalatable polyunsaturated fats to the diet. For many Americans this would be unrealistic. Fiber is another good examples. One could expect, at the most, a 5% decrease in serum cholesterol by adding any reasonable amount of fiber to the diet.

Recent studies confirming that reduction of elevated serum cholesterol levels can reduce risk of heart disease have rekindled interest in the effects of diet on serum cholesterol levels and other risk factors for heart disease. We have known for years that one of the most important dietary factors regulating serum cholesterol levels is the ratio of polyunsaturated fats (PUFAs) to saturated fats (SFAs) in the diet. One of the most interesting recent developments is the discovery that different types of polyunsaturated fatty acids have different effects on lipid metabolism and on other risk factors for heart disease. There are two families of polyunsaturated essential fatty acids— the ω-6, or linolenic family, and the ω-3, or linolenic family. Recent clinical studies have shown that the ω-6 PUFAs (chief dietary source is linolenic acid from plants and vegetable oils) primarily decrease serum cholesterol levels, with only modest effects on triacylglycrol levels. The ω-3 PUFAs (chief dietary source is eicosapentaenoic acid from certain ocean fish and fish oils) cause modest decreases in serum cholesterol levels and significantly lower triacylglycerol levels. The biochemical mechanism behind these different effects on serum lipid levels is unknown.

The ω-3 PUFAs have yet another unique physiological effect that may decrease the risk of heart disease—they decrease platelet aggregation. The mechanism of this effect is a little clearer. Arachidonic acid (ω-6 family) is known to be a precursor of thromboxane A_2 (TXA_2), which is a potent proaggregating agent, and prostaglandin I_2 (PGI_2), which is a weak antiaggregating agent. The ω-3 PUFAs are thought to act by one of two mechanisms: (1) Eicosapentaenoic acid (ω-3 family) may be converted to thromboxane A_3 (TXA_3), which is only weakly proaggregating, and prostaglandin I_3 (PGI_3), which is strongly antiaggregating. Thus the balance between proaggregation and antiaggregation would be shifted toward a more antiaggregating condition as the ω-3 PUFAs displace ω-6 PUFAs as a source of precursors to the thromboxanes and prostaglandins. (2) The ω-3 PUFAs may also act by simply inhibiting the conversion of arachidonic acid to TXA_2.

The unique potential of eicosapentaenoic acid and other ω-3 PUFAs in reducing the risk of heart disease is being tested in numerous clinical trials. Although the results may affect dietary recommendations in the future, it is well to keep in mind that no long-term clinical studies of the ω-3 PUFAs have been carried out. No major health organization has recommended that we replace ω-6 with ω-3 PUFAs in the American diet.

Iron Metabolism

14.1 Introduction

Iron is closely involved in the metabolism of oxygen, permitting the transportation and participation of oxygen in a variety of biochemical processes. The common **oxidation states** are either ferrous (Fe^{2+}) or **ferric** (Fe^{3+}); higher oxidation levels occur as short-lived intermediates in certain redox processes. Iron has an affinity for electronegative atoms such as oxygen, nitrogen, and sulfur, which provide the electrons that form the bonds with iron. These can be of very high affinity when favorably oriented on macromolecules. In forming complexes, no bonding electrons are derived from iron. There is an added complexity to the structure of iron; the nonbonding electrons in the outer shell of the metal (the incompletely filled 3*d* orbitals) can exist in two states. Where bonding interactions with iron are weak, the outer nonbonding electrons will avoid pairing and distribute throughout the 3*d* orbitals. Where bonding electrons interact strongly with iron, however, there will be pairing of the outer nonbonding electrons, favouring lower-energy 3*d* orbitals. These two different distributions for each oxidation state of iron can be determined by electron spin resonance measurements. Dispersion of 3*d* electrons to all orbitals leads to the high-spin state, whereas restriction of 3*d* electrons to lower energy orbitals, because of electron pairing, leads to a low-spin state. Some iron-protein complexes reveal changes in spin state without changes in oxidation during chemical events (e.g., binding and release of oxygen by hemoglobin).

At neutral and alkaline pH ranges, the redox potential for iron in aqueous solutions favors the Fe^{3+} state; at acid pH values, the equilibrium favours the Fe^{2+} state. In the Fe^{3+} state ion slowly forms large polynuclear complexes with hydroxide iron, water, and other anions that may be present. These complexes can become so large as to exceed their solubility products, leading to their aggregation and precipitation with pathological consequences.

Iron can bind to and influence the structure and function of various macromolecules, with deleterious results to the organism. To protect against such reactions, several iron-binding proteins function specifically to store and transport iron. These proteins have both a very high affinity for the metal and, in the normal physiological state, also have incompletely filled iron-binding sites. The interaction of iron with its ligands has been well characterized in some proteins (e.g., hemoglobin and myoglobin), whereas for others (e.g., transferrin) it is presently in the process of being defined. The major area of ignorance in the biochemistry of iron lies in the *in vivo* transfer processes of iron from one macromolecule to another. Several proposed mechanisms may explain the process of iron transfer. Two are supported by excellent model studies but have varying degrees of relevance to the physiological state. The proposed processes are the following. First, the redox change of iron has been an attractive mechanism because it is supported by selective *in vitro* studies and because in some cases macromolecules have a very selective affinity for Fe^{3+}, binding Fe^{2+} poorly. Thus reduction of iron would permit ferrous ions to dissociate, and reoxidation would allow the iron to redistribute to appropriate macromolecules. Redox mechanisms have only been defined in a very few settings, some of which will be described below. An alternative hypothesis involves chelation of ferric ions by specific small molecules with high affinities for iron; this mechanism has been supported also by selective *in vitro* studies. The chelation mechanism suffers from the lack of a demonstrably specific *in vivo* chelator. Because the redox potential strongly favours ferric ion at almost all tissue sites and because Fe^{3+} binds so strongly to liganding groups, the probability is that there are cooperating mechanisms regulating the intermolecular transfer of iron.

Iron binds to proteins either by incorporation into a **protoporphyrin IX** ring or by interaction with other protein ligands. Ferrous- and ferric-protoporphyrin IX complexes are designated **heme** and **hematin,** respectively. Heme-containing proteins include those that transport (e.g., hemoglobin) and store (e.g., myoglobin) oxygen, and certain enzymes that contain heme as part of their prosthetic groups (e.g., catalase, peroxidases, tryptophan pyrrolase, prostaglandin synthase, guanylate cyclase, NO synthase, and the microsomal and mitochondrial cytochromes.).

Nonheme proteins include transferrin, ferritin, a variety of redox enzymes that contain iron at the active site, and iron-sulfur proteins. A significant body of information has been acquired that relates to the structure-function relationships of some of these molecules.

If an individual is overloaded with iron by any of several causes, the serum transferrin value can be close to saturation, making small amounts of free serum iron available. Microorganisms that are usually nonpathogenic, because they are iron dependent and cannot compete against partially saturated transferrin in the normal individual, can now become pathogenic under these circumstances. For example, *Vibrio vulnificus,* a marine halophile, is found in a small percentage of oysters and commercial shellfish. Individuals who are iron overloaded can develop a rapidly progressive infection, with death ensuing within 24 h after ingestion of the offending meal, whereas normal individuals consuming the same food are entirely free of symptoms.

14.2 Iron-Responsive Element

Cytosol contains at least two proteins that respond to changes in iron concentration. They act as effector molecules controlling the translation of mRNAs, which are important in iron metabolism. These **iron regulatory proteins (IRPs)** bind to specific **stem-loop structures** on certain mRNAs. IRP-1 is the best defined of these proteins. It contains an Fe_4S_4 cubane group when the cellular concentration of iron is high. This prosthetic group activates IRP-1 so that in possesses an **aconitase** activity. However, since neither citrate nor isocitrate is present in significant amounts in the cytosol, the activity is only a potential one.

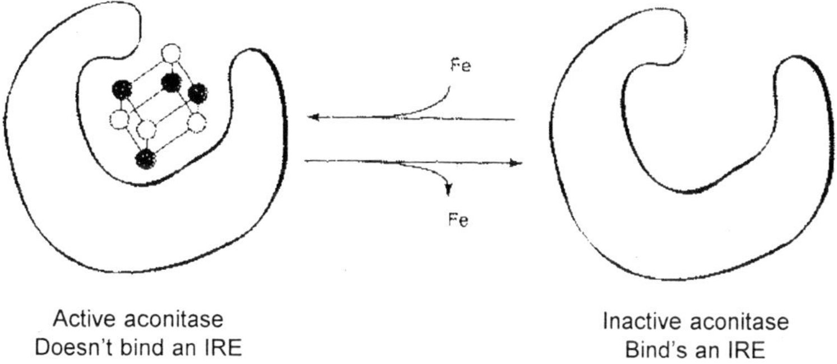

Iron Responsive Potential-I.

At low iron concentrations, the cubane structure collapses, dissociating from the protein and leaving an apoenzyme without catalytic activity. However, it can now bind to specific mRNA stem-loop structures, known as **iron-responsive elements (IREs).** Five mRNAs are known to contain IREs: those for the light and heavy chains of ferritin, the erythrocytic form of aminolevulinic acid synthase, the mitochondrial form of aconitase, and transferrin receptor. (Mitochondrial aconitase, the physiologically active isozyme, has no IRP function.) The first four mRNAs have single IREs in the 5′ flanking region, which bind a single IRP. In contrast, the transferrin receptor has five tandem IREs that bind IRPs in the 3′ flanking region. The binding of the 5′ and 3′ flanking IREs leads to different translational effects. In the iron-deprived state, binding to the 3′ IRE of transferrin receptor leads to stabilization of the mRNA with reduced turnover and, therefore, an increased number of receptor-specific RNA molecules, thereby leading to the increased synthesis of receptor protein. The single 5′ stem-loop of ferritin mRNA is homologous to the 3′ stem-loops of the transferrin receptor mRNA. However, in the former case, binding of the IRP leads to a decreased rate of translation of the mRNA and, thereby, to a decreased concentration of ferritin molecules. Note that the molecular events that are controlled are different in the syntheses of transferrin receptor and apoferritin.

In summary, low iron concentrations lead to activation of an IRP that binds to the mRNAs for transferrin receptor and ferritin. In the former case, more receptor is synthesized, while in the latter case less apoferritin is synthesized. The net effect is utilization of iron by proliferating cells. In contrast, high iron concentrations lead to loss of binding by the IRPs to IREs, with a shift of iron from uptake by proliferating cells to storage in the liver.

IRP-1 is regulated by its change from active to inactive states in mRNA-binding properties as noted above **IRP-2**, a second regulatory protein, also responds to varying concentrations of iron, but in this case, the protein is regulated by increased synthesis at low iron concentrations and increased degradation by a proteasome at high iron concentrations. In addition to the effects of changed iron concentration, increased production of NO also acts to regulate IRPs.

Single mutations have been described of two adjacent bases in the loop segment of the iron-responsive element of ferritin light chain mRNA with an increased amount of apoferritin being synthesized but without an increase in total body iron. This mutation leads to a 28-fold lower affinity for IRP-1 in one case and perhaps an even lower affinity in the other. The reason why these patients have cataracts is unknown. The gene for MP-19, an abundant protein in the lens, which is very close to the light chain gene on chromosome 19, might possibly be affected by the regulatory process on the mRNA. However, it is more probable that a greatly increased synthesis of ferritin in the lens leads to an increased amount of iron-catalyzed reactions with well-described oxidative lenticular damage.

14.3 Biosynthesis of Bilirubin

Bilirubin is derived not only from senescent red cells but also from the turnover of other heme-containing proteins, such as the cytochromes. Studies will labeled glycine as a precursor have revealed that an early-labeled bilirubin, with a peak within 1-3 h, appears a very short time after a pulsed administration of the labeled precursor. A larger amount of bilirubin appears much later at about 120 days, reflecting the turnover of heme in red blood cells. Early-labeled bilirubin can be divided into two parts: an early-early part, which reflects the turnover of heme proteins in the liver, and a late-early part, which consists of both the turnover of heme-containing hepatic proteins and the turnover of bone marrow heme, which is either poorly incorproated or easily released from red blood cells. The latter is a measurement of ineffective erythropoiesis and can be very pronounced in disease states such as pernicious anemia and the thalassemias.

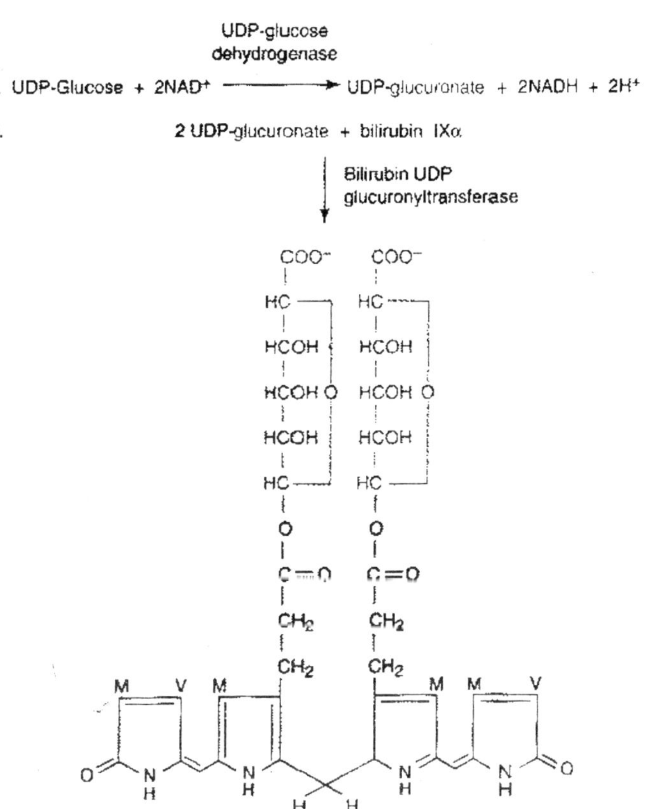

Biosynthesis of bilirubin diglucuronide.

Bilirubin is poorly soluble in aqueous solutions at physiological pH values. When transported in plasma, it is bound to serum albumin with an association constant greater than 10^6 M^{-1}. Albumin contains one such high-affinity site and another with a lesser affinity. At the normal albumin concentration of 4 g dL^{-1}, about 70 mg of bilirubin per deciliter of plasma can be bound on the two sites. However, bilirubin toxicity (**kernicterus**), which is manifested by the transfer of bilirubin to membrane lipids, commonly occurs at concentrations greater than 25 mg dL^{-1}. This suggests that the weak affinity of the second site does not allow it to serve effectively in the transport of bilirubin. Bilirubin on serum albumin is rapidly cleared by the liver, where there is a free bidirectional flux of the tetrapyrrole across the sinusoidal-hepatocyte interface. Once in the hepatocyte, bilirubin is bound to several cytostolic proteins, of which only one has been well characterized. The latter component, **ligandin,** is a small basic component making up to 6% of the total cytosolic protein of rat liver. Ligandin has been purified to homogeneity from rat liver and characterized as having two subunits with molecular masses of 22 kDa and 127 kDa. Each subunit contains glutathione *S*-epoxidetransferase activity, a function important in detoxification mechanisms of aryl groups. The stoichiometry of binding is one bilirubin molecule per complete ligandin molecule. The functional role of ligandin and other hepatic bilirubin-binding proteins remains to be defined.

Once in the hepatocyte the propionyl side chains of bilirubin are conjugated to form a diglucuronide. The reaction utilizes uridine diphosphoglucuronate derived from the oxidation of uridine diphosphoglucose. The former serves as a glucuronate donor to bilirubin. In normal bile, the diglucuronide is the major form of excreted bilirubin, with only small amounts of the monoglucuronide or other glycosidic adducts present. **Bilirubin diglucuronide** is much more water-soluble than free bilirubin, and thus the transferase facilitates excretion of the bilirubin into bile. Bilirubin diglucuronide is poorly absorbed by the intestinal mucosa. The glucuronide residues are released in the terminal ileum and large intestine by bacterial hydrolases; the released free bilirubin is reduced to the colorless linear tetrapyrroles known as **urobilinogens.** Urobilinogens can be oxidized to coloured products known as **urobilins,** which are excreted in the feces. A small fraction of urobilinogen can be reabsorbed by the terminal ileum and large intestine to be removed by hepatic cells and resecreted in bile. When urobilinogen is reabsorbed in large amounts in certain disease states, the kidney serves as a major excretory site.

In the normal state, plasma bilirubin concentrations are 0.3-1 mg dL^{-1}, and this is almost all in the unconjugated state. In the clinical setting, conjugated bilirubin is expressed as **direct bilirubin** because it can be coupled readily with diazonium salts to yield azo dyes; this is the direct **van den Bergh reaction.** Unconjugated bilirubin is bound noncovalently to albumin and will not react until it is released by the addition of an organic solvent such as ethanol. The reaction with diazonium salts yielding the azo dye after the addition of ethanol is the indirect van den Bergh reaction, and this measures the **indirect bilirubin** or the unconjugated bilirubin. Unconjugated bilirubin binds so tightly to serum albumin and lipid that it does not diffuse freely in plasma and therefore does not lead to an elevation of bilirubin in the urine. Unconjugated bilirubin has a high affinity for membrane lipids, which leads to the impairment of cell membrane function, especially in the nervous system. In contrast, conjugated bilirubin is relatively water-soluble, and elevations of this bilirubin form lead to high urinary concentrations with the characteristic deep yellow-brown colour. The deposition of conjugated and unconjugated bilirubin in skin and the sclera gives the yellow to yellow-green colour seen in patients with jaundice.

A third form of plasma bilirubin occurs only with hepatocellular disease in which a fraction of the bilirubin binds so tightly that it is not released from serum albumin by the usual techniques and is linked covalently to the protein. In some cases up to 90% of total bilirubin can be in this covalently bound form.

The normal liver has a very large capacity to conjugate and mobilize the bilirubin that is delivered. As a consequence, **hyperbilirubinemia** due to excess heme destruction, as in hemolytic diseases, rarely leads to bilirubin levels that exceed 5 mg dL^{-1}, except in situations in which functional derangement of the liver is present. Thus marked elevation of unconjugated bilirubin reflects primarily a variety of hepatic diseases, including those that are heritable and those that are acquired.

Elevations of conjugated bilirubin level in plasma are attributable to liver and/or biliary tract disease. In simple uncomplicated biliary tract obstruction, the major component of the elevated serum bilirubin is the diglucuronide form, which is released by the liver into the vascular compartment.